5 STEPS TO A 5™

AP Physics 1: Algebra-Based

2016

5 STEPS TO A 5™

AP Physics 1: Algebra-Based

2016

Greg Jacobs

Mc
Graw
Hill
Education

New York   Chicago   San Francisco   Athens   London   Madrid
Mexico City   Milan   New Delhi   Singapore   Sydney   Toronto

2 3 4 5 6 7 8 9 0   RHR/RHR   1 2 1 0 9 8 7 6 5

ISBN   978-0-07-184639-4
MHID   0-07-184639-5
ISSN   2334-5349

e-ISBN   978-0-07-184640-0
e-MHID   0-07-184640-9
ISSN   2334-5373

McGraw-Hill Education, the McGraw-Hill Education logo, 5 Steps to a 5, and related trade dress are trademarks or registered trademarks of McGraw-Hill Education and/or its affiliates in the United States and other countries and may not be used without written permission. All other trademarks are the property of their respective owners. McGraw-Hill Education is not associated with any product or vendor mentioned in this book.

AP, Advanced Placement Program, and College Board are registered trademarks of the College Entrance Examination Board, which was not involved in the production of, and does not endorse, this product.

The series editor was Grace Freedson, and the project editor was Del Franz.

Series design by Jane Tenenbaum.

McGraw-Hill Education products are available at special quantity discounts to use as premiums and sales promotions or for use in corporate training programs. To contact a representative, please visit the Contact Us pages at www.mhprofessional.com.

AP Teachers: Order your free online Teacher's Manual
with teaching strategies, student activity and project ideas, and other ways to incorporate the review materials and practice tests in this *5 Steps to a 5* guide into your classroom curriculum.

Download your free Teacher's Manual from:

http://www.mhprofessional.com/promo/5steps/

or scan this QR code:

ABOUT THE AUTHOR

Greg Jacobs is chairman of the science department at Woodberry Forest School, the nation's premier boarding school for boys. Over the years, Greg has taught all flavors of AP physics. He is a reader and consultant for the College Board, which means he grades AP physics exams, and he runs professional development seminars for other AP teachers. Greg is president of the USAYPT, a nonprofit organization promoting physics research at the high school level. Greg was recently honored as an AP Teacher of the Year by the Siemens Foundation. Outside the classroom, Greg has coached football, baseball, and debate. He is the play-by-play voice of Woodberry sports on the Internet, calling football, baseball, soccer, and basketball games. Greg writes the prominent physics teaching blog available at www.jacobsphysics.blogspot.com.

CONTENTS

Preface, xi

Acknowledgments, xiii

Introduction: The Five-Step Program, xv

STEP 1   Get to Know the Exam and Set Up Your Study Program

1 **Frequently Asked Questions About the AP Physics 1 Exam 3**
FAQs: The AP Physics Program 4
FAQs: The AP Physics 1 Exam 6

2 **Understanding the Exam: The AP Physics 1 Revolution 13**
What Happened to the AP Physics Test? 14
What Is AP Physics 1? Eleven Things You Should Know About the Course and
Exam 14

3 **How to Use Your Time 21**
Personalizing Your Study Plan 21
Plan A: You Have a Full School Year to Prepare 22
Plan B: You Have One Semester to Prepare 23
Plan C: You Have Six Weeks to Prepare 23

STEP 2   Determine Your Test Readiness

4 **Test Yourself: AP Physics 1 Fundamentals 27**
Self-Assessment: AP Physics 1 Fundamentals 27
Solutions for the AP Physics 1 Fundamentals Self-Assessment 29

5 **Test Yourself: AP Physics 1 Question Types 33**
Self-Assessment: Question Types 33
Solutions for the AP Physics 1 Question Types Assessment 42

STEP 3   Develop Strategies for Success

6 **Strategies to Get the Most Out of Your AP Physics Course 49**
Seven Simple Strategies to Get the Most Out of Your AP Physics Course 49

7 **Strategies to Approach the Questions on the Exam 53**
Tools You Can Use and Strategies for Using Them 54
Strategies for Questions That Involve a Ranking Task 55
Strategies for Questions That Involve Graphs 56

8 **Strategies to Approach the Questions: Free-Response Section 61**
Structure of the Free-Response Section 62
How to Approach the Laboratory Question 62
The Qualitative-Quantitative Translation (QQT) 65
What Do the Exam Readers Look For? 67
Final Advice About the Free-Response Questions 69

9 **Strategies to Approach the Questions: Multiple-Choice Section 71**
Multiple-Choice Questions 72
Multiple-Correct: A New Question Type 72

Preparing for the Multiple-Choice Section of the Test 73
Final Strategies for the Multiple-Choice Section 73

STEP 4   **Review the Knowledge You Need to Score High**

10 Motion in a Straight Line 77
Introduction to Motion in a Straight Line 78
Graphical Analysis of Motion 78
Algebraic Analysis of Motion 81
Practice Problems 85
Solutions to Practice Problems 86
Rapid Review 87

11 Forces and Newton's Laws 89
Describing Forces: Free-Body Diagrams 90
Determining the Net Force 92
Newton's Third Law 93
Forces at Angles 93
Practice Problems 97
Solutions to Practice Problems 98
Rapid Review 99

12 Collisions: Impulse and Momentum 101
The Impulse-Momentum Theorem 102
Conservation of Momentum 103
Motion of the Center of Mass 108
Practice Problems 108
Solutions to Practice Problems 109
Rapid Review 111

13 Work and Energy 113
Energy 114
Work 115
The Work-Energy Theorem 117
Power 118
Practice Problems 119
Solutions to Practice Problems 120
Rapid Review 121

14 Rotation 123
Circular Motion 124
Torque 126
Rotational Kinematics 128
Rotational Inertia 129
Newton's Second Law for Rotation 130
Angular Momentum 131
Rotational Kinetic Energy 134
Practice Problems 135
Solutions to Practice Problems 135
Rapid Review 137

15 Gravitation 139
Determining the Gravitational Field 140
Determining Gravitational Force 141

Force of Two Planets on One Another—Order of Magnitude Estimates 141
Gravitational Potential Energy 142
Gravitational and Inertial Mass 143
Fundamental Forces: Gravity Versus Electricity 143
Practice Problems 144
Solutions to Practice Problems 145
Rapid Review 147

16 Electricity: Coulomb's Law and Circuits 149
Electric Charge 150
Circuits 152
Practice Problems 160
Solutions to Practice Problems 161
Rapid Review 162

17 Waves and Simple Harmonic Motion 165
Simple Harmonic Motion 166
Waves 169
Practice Problems 176
Solutions to Practice Problems 177
Rapid Review 178

18 Extra Drills on Difficult but Frequently Tested Topics 181
How to Use This Chapter 181
Springs and Graphs 182
Tension 188
Inclined Planes 191
Motion Graphs 194
Simple Circuits 198

STEP 5 **Build Your Test-Taking Confidence**
AP Physics 1 Practice Exam 1: Section I (Multiple-Choice), 205
AP Physics 1 Practice Exam 1: Section II (Free-Response), 219
Solutions: AP Physics 1 Practice Exam 1, Section I (Multiple-Choice), 223
Solutions: AP Physics 1 Practice Exam 1, Section II (Free-Response), 229
AP Physics 1 Practice Exam 2: Section I (Multiple-Choice), 235
AP Physics 1 Practice Exam 2: Section II (Free-Response), 251
Solutions: AP Physics 1 Practice Exam 2, Section I (Multiple-Choice), 255
Solutions: AP Physics 1 Practice Exam 2, Section II (Free-Response), 261
Scoring the Practice Exams, 267

Appendixes
Table of Information, 271
The Pantheon of Pizza, 275

Why is this book different from all other AP Physics prep books?

The quality of the prep book starts with the author. Greg Jacobs has taught all versions of AP (and not-AP) physics since 1996. His hundreds of students over the years have a greater than 99 percent pass rate; more than 70 percent earn 5s. Greg has graded the exams, he has written questions that have appeared on exams, and he has taught teachers how to teach to the exams. The leading blog about physics teaching is his—"Jacobs Physics." (Take a look at it.) Your author isn't some med student looking for extra cash, or a no-name college professor. Greg knows what he's talking about, and the proof is in the results and feedback from his students.

AP Physics 1 emphasizes conceptual understanding over algebraic manipulation and so does this book. So many people presume that physics is about finding the right numbers to plug into the right equation. That couldn't be farther from the truth. Successful physics students can explain why they chose a specific equation. They can explain what values for each variable are reasonable, and why. They can explain the *physical* meaning of any mathematical manipulation—how would this problem look in the laboratory? What equipment would be used to measure these values? The AP Physics 1 exam mostly asks questions that are not about number crunching. Greg has been teaching "beyond the numbers" for years, and he brings his expertise in explaining complex concepts in simple language.

Your textbook is impenetrable, even to senior physics majors. This prep book is readable. Be honest—when you read your textbook, you really just try the problems at the end of the chapter, then look back for a template of how to do those problems. Well, this book's content review is structured around this very method. Greg poses example equations and talks you through them. On the way, he shows you the relevant facts and equations, as well as *how you are supposed to know they are relevant*. He explains not just the answer, but the thought process behind the answer. You do not have to be already good at physics to understand the text.

This book's practice tests and practice questions are authentic. One of the primary tenets of Greg's physics teaching is that in-class tests should look exactly like the real AP exams. The College Board has published a curriculum guide that provides express guidance as to the style and content of the exam questions. Greg's practice tests are derived directly from what the curriculum guide says. He has vast experience phrasing questions in the style seen on College Board exams.

Every practice question includes not just an answer, but also a thorough explanation of how to get that answer. The back of your textbook may provide answers like "2 m/s" or "increase." Huh? Skim through and look at the solutions to the practice exam, and to the end-of-chapter questions. They're complete. This book's solutions explain everything, even sometimes the common mistakes that you might have made by accident. Exam readers expect thoroughness on the free-response problems—you should expect the same from your prep book.

The AP Physics 1 Exam is still somewhat mysterious because it's only been administered once. This book, like all other prep books, goes to press well before the release of the first official exam. So don't believe any prep book that says it knows exactly what the new exam is going to be like. The only people who truly know everything about the Physics 1 exam are those on the test development committee and the Educational Testing Service (ETS) professionals who are helping the College Board create the exam. They aren't talking. All a prep book author can do is read the curriculum guide and the released practice exam, talk in generalities with official College Board representatives, and make a best guess. There's no doubt that, while Greg will be right-on about most things he says in this book, he'll have to change a few things once several released exams show patterns. But just as Kirk trusted Spock's guesses more than most people's sureties, you should trust Greg's instincts just as hundreds of students have over the decades.

ACKNOWLEDGMENTS

The original idea for this book came many moons ago, when my AP students in 1999 couldn't stand the other review books on the market. "We can write a better book than that," they said. And we did. Justin Kreindel, Zack Menegakis, Adam Ohren (who still owes me four sandwiches from Mississippi Sweets for poking Jason during class), Jason Sheikh, and Joe Thistle were the cast members for that particular opera.

Josh Schulman, also a member of that 1999 class, was the one who buckled down and put pen to paper, or rather fingers to computer keys. His first draft of a book forced me to revise and finish and publish. Josh is still a coauthor on McGraw-Hill's *5 Steps to a 5: AP Physics C*. I highly recommend that book.

Del Franz has been a fabulous editor for this and other projects. I appreciate his toils.

The faculty and administration at Woodberry Forest School, in particular former science department chairman **Jim Reid**, deserve mention. They have been extraordinarily supportive of me professionally.

Two amazing physics teachers have vetted this new book for the new exam. They have done tremendous work in exchange for nothing but a free cup of coffee. Thank you, **Jeff Steele** (who gave feedback on the practice tests) and **Matt Sckalor** (who read every content chapter). I owe you both.

Thank you to those who sent in errata from the first edition. I'm particularly indebted to Glenn Mangold and Drew Austen, who sent in three pages worth of edits, along with careful and thorough justifications for each—wow. Joseph Rao, Patrick Diehl, Mike Pozuc, Shannon Copeland, and I'm sure others I've missed have also sent in corrections or thoughts that I've used. And to all of the Jacobs Physics blog readers, know that I deeply appreciate the comments, e-mails, and conversations you send my way.

Most important, thank you to Shari and Milo for putting up with me during all of my writing projects.

INTRODUCTION: THE FIVE-STEP PROGRAM

Welcome!

I know that preparing for the Advanced Placement (AP) Physics 1, Algebra-Based Exam can seem like a daunting task. There's a lot of material to learn and some of it can be challenging. But I also know that preparing for the AP exam is much easier—and a lot more enjoyable—if you do it with a friendly, helpful guide. So order the pizza (see Pantheon of Pizza in the Appendix) and let's get started.

First, you should know that physics does not lend itself well to cramming. Success on the AP exam is most likely the result of actually learning and *understanding* physics in your AP Physics 1 course. If you are opening this book in the first semester or early in the second semester, be sure to read Chapter 6, which contains strategies to get the most out of your physics class in terms of preparation for the AP exam.

Of course, this book can also be instrumental in helping you score high. *5 Steps to 5: AP Physics 1, Algebra-Based* is composed of practical, score-raising items you won't necessarily get in your AP course, including in-depth information about the test, proven strategies to attack each type of question on the exam, an easy-to-follow review of content, and a very realistic practice test.

Organization of the Book: The Five-Step Program

You will be taking a lengthy, comprehensive exam in May. You want to be well prepared so that the exam takes on the feel of a command performance, not a trial by fire. Following the Five-Step program is the best way to structure your preparation.

Step 1: Get to Know the Exam and Set Up Your Study Program

You need to get to know the exam—what's on it and how it's structured—so there are no surprises on test day. Understanding the test is the first step in preparing for it. And you need a plan. Step 1 gives you the background and structure you will need before you start exam preparation.

Step 2: Determine Your Test Readiness

Your study program should *not* include cramming absolutely everything about physics into your head in the weeks before the test; it can't be done. Instead, you'll need to assess your strengths and weaknesses and prioritize what you need to review. The physics fundamentals self-assessment in Chapter 4 will help you do just that. Note that the questions in this self-assessment are not written in the style of the actual questions on the AP Physics 1 Exam. They are designed to quickly determine your strengths and weakness, not to mimic actual test questions.

Then, in Chapter 5 you'll be introduced to the different types of questions found on the actual AP Physics 1 Exam. The self-assessment in this chapter allows you to see how you do on each of the different types of questions and to identify the question types with which you need the most practice. The results from both self-assessments—the fundamentals self-assessment in Chapter 4 and the question-type assessment in Chapter 5—should help you develop your study plan and determine which chapters in this book you'll spend the most time on.

Step 3: Develop Strategies for Success

First, as mentioned in preceding text, Chapter 6 contains strategies to get the most out of your AP Physics 1 course in terms of being able to get a high score on the exam. Read this if it's the first semester or beginning of the second semester of your course.

The focus of the remaining chapters in Step 3 is developing effective strategies to approach each of the question types found on the AP Physics 1 Exam. Sure, I know you've been listening to general test-taking advice and have been taking multiple-choice standardized tests practically your whole life. But the chapters in this section contain *physics*-specific advice. And the new AP Physics 1, Algebra-Based Exam has questions—even in the multiple-choice section—that are probably unlike any you've encountered previously in your standardized test taking. Chapter 7 focuses on strategies for question-types found on both the multiple-choice and free-response sections of the test. Chapter 8 looks at strategies for types of questions found only on the free-response section, and finally, Chapter 9 suggests strategies for question types that appear only in the multiple-choice section of the test.

Be aware that knowing physics in itself won't automatically give you the best test score you are capable of getting. To do your best, you'll need to understand the different kinds of questions on the test and develop the most effective strategy for attacking each question type.

Step 4: Review the Knowledge You Need to Score High

Step 4 contains a comprehensive review of the topics on the AP exam. Now, you've probably been in an AP Physics class all year and you've likely read (or at least tried to read!) your textbook. Our review is not meant to be another textbook. It's only a review—an easy-to-follow, step-by-step review focused exclusively on the things likely to appear on the AP exam. The review is not as detailed as your textbook, but it's more germane to what's actually on the AP Physics 1 Exam. Chapters 10 through 15 provide a review of different aspect of mechanics, Chapter 16 focuses on electricity, and Chapter 17 focuses on waves.

These review chapters are appropriate both for quick skimming (to remind yourself of salient points) and for in-depth study, working through each sample problem. Each review chapter contains several questions in the format of the free-response questions actually found on the test. Use these questions both to test your knowledge and to practice with the types of questions you'll encounter on the test.

Finally, in Chapter 18, you'll find extra drills on some of the most common physics situations tested on the AP exam. The old saying is true: practice makes perfect.

Step 5: Build Your Test-Taking Confidence

This is probably the most important part of this book: full-length practice tests that closely reflect what you'll encounter in the actual test. Unlike other practice tests you may take, these come with thorough explanations. One of the most important elements in learning physics is making, and then learning from, mistakes. This book doesn't just tell you what you got wrong; we explain why your answer is wrong and how to do the problem correctly. It's okay to make a mistake here because, if you do, you probably won't make the same mistake again on that day in mid-May. In fact, it's a good idea to read not only the solutions to the problems you got wrong, but also the solutions for the problems you weren't sure of or simply happened to guess correctly.

The Graphics Used in This Book

To emphasize particular skills and strategies, we use icons throughout this book. An icon in the margin will alert you that you should pay particular attention to the accompanying text. We use these three icons:

1. This icon points out a very important concept or fact that you should not pass over.

2. This icon calls your attention to a problem-solving strategy that you may want to try.

3. This icon indicates a tip that you might find useful.

5 STEPS TO A 5™

AP Physics 1: Algebra-Based

2016

STEP 1

Get to Know the Exam and Set Up Your Study Program

CHAPTER **1** Frequently Asked Questions About the AP Physics 1 Exam

CHAPTER **2** Understanding the Exam: The AP Physics 1 Revolution

CHAPTER **3** How to Use Your Time

CHAPTER 1

Frequently Asked Questions About the AP Physics 1 Exam

IN THIS CHAPTER

Summary: This chapter provides the basic information you need to know about the AP Physics 1, Algebra-Based Exam. Learn how the test is structured, what topics are tested, how the test is scored, as well as basic test-taking information.

Key Ideas

✪ It's not possible to "game" this test. In order to get a good score, *you must know your physics*.

✪ Half of the test consists of multiple-choice questions and the other half of free-response questions. Each section accounts for half of your score.

✪ Most colleges and universities will award credit for scoring a 4 or a 5 on the exam. Some schools even accept a score of 3 on the exam.

✪ Topics on the exam include kinematics; forces; gravitation; impulse-momentum; energy; rotation, motion, torque, and angular momentum; electricity; and mechanical waves, sound, and simple harmonic motion.

✪ The focus of the test is not numbers and equations. You may use a calculator and an equation sheet, but these will not be very helpful because far more explanations and verbal responses are required than calculations and numerical answers.

FAQs: The AP Physics Program

This chapter contains the answers to some of the most frequently asked questions about the AP Physics 1, Algebra-Based course and exam. If you have additional questions, check out the College Board's "AP Central" web pages (http://apcentral.collegeboard.com). Another helpful resource for the test is the author's physics teaching blog at http://jacobsphysics .blogspot.com.

What Is AP Physics 1, Algebra-Based, and How Is It Different from a Typical Advanced Physics Course?

AP Physics 1 is a first-time, no-calculus physics course covering mechanics, waves, and electricity.[1] The AP Physics 1 exam involves fewer topics than typical high school or college introductory courses, but it requires far more explanations and verbal responses than calculations and numerical answers.

Even though most advanced physics courses require loads of numerical answers and mathematical manipulation, AP Physics 1 requires you to be able to do only to things mathematically: (1) solve straightforward algebraic equations, and (2) use the basic definitions of the trigonometric functions sine, cosine, and tangent. There's no completing the square, no trigonometric identities—just the basic stuff you learned in your algebra and geometry courses.

The next chapter contains more information about how the AP Physics 1, Algebra-Based curriculum differs from the old AP Physics B course and other traditional advanced physics courses.

Who Should Take the AP Physics 1, Algebra-Based Course?

The Physics 1 course is ideal for *all* college-bound, high school students. For those who intend to major in math or the heavy-duty sciences, Physics 1 serves as a perfect introduction to college-level work. For those who want nothing to do with physics after high school, Physics 1 is a terrific terminal course—you get exposure to many facets of physics at a rigorous, yet understandable level.

Most important, for those who aren't sure in which direction their college career may head,[2] the Physics 1 course can help you decide: "Do I like this stuff enough to keep studying it, or not?"

---

[1]The first AP Physics 1 exam was given in May 2015. For the previous four decades, AP Physics B was the College Board's algebra-based introductory physics exam.

[2]That may be most of you reading this book.

What Are the Other AP Physics Courses?

In addition to AP Physics 1, the College Board now offers three other AP Physics courses.

AP Physics 2 is designed as an algebra-based follow-up to AP Physics 1. In the same style of requiring depth of understanding and verbal explanation, AP Physics 2 covers electricity, magnetism, fluids, thermal physics, atomic and nuclear physics, and more.

The AP Physics C courses are *only* for those who have already taken a solid introductory physics course and are considering a career in physical science or math. Physics C consists of two separate, calculus-based courses: (1) Newtonian Mechanics, and (2) Electricity and Magnetism. Of course, the Physics 1 and Physics 2 courses cover these topics as well. However, the C courses go into greater mathematical depth and detail. The problems are more involved, and they demand a higher level of conceptual and mathematical ability, including differential and integral calculus, and some differential equations. You can take either or both 90-minute Physics C exams. The AP Physics C exams have not changed in many years. If you decide to attempt the Physics C Exam, try *5 Steps to a 5: AP Physics C*.

Is One Exam Better than the Other? Should I Take More than One?

We strongly recommend taking only one exam—and make sure it's the one your high school AP course prepared you for! Physics C is not considered "better" than Physics 1 or 2 in the eyes of colleges and scholarship committees. They are different courses with different intended audiences. It is far better to do well on the one exam you prepared for than to attempt something else and do poorly.

Why Should I Take an AP Physics Exam?

Many of you take the AP Physics Exam because you are seeking college credit. The majority of colleges and universities will award you some sort of credit for scoring a 4 or a 5. A smaller number of schools will even accept a score of 3 on the exam. This means you are one or two courses closer to graduation before you even start college!

Therefore, one compelling reason to take the AP Exam is economic. How much does a college course cost, even at a relatively inexpensive school? You're talking several thousands of dollars. If you can save those thousands of dollars by paying less than a hundred dollars now, why not do so? Even if you do not score high enough to earn college credit, the fact that you elected to enroll in an AP course tells admissions committees that you are a high achiever and are serious about your education. In recent years, about 60 percent of students taking an AP Physics exam have scored 3 or higher. Your odds of success are good.

You'll hear a whole lot of misinformation about AP credit policies. Don't believe anything a friend (or even a teacher) tells you; instead, find out for yourself. One way to learn about the AP credit policy of the school you're interested in is to look it up on the College Board's official website, at http://collegesearch.collegeboard.com/apcreditpolicy/index.jsp. Even better, contact the registrar's office or the physics department chair at the college directly.

FAQs: The AP Physics 1 Exam

This Is a New Exam: How Am I Supposed to Know What to Expect?

The first AP Physics 1 exam was given in May 2015. The free response questions from that exam are available on the collegeboard.com website; however, the multiple choice questions will not be released. The College Board has published an enormous amount of background material for this new exam on their official AP website, "AP Central" (http://apcentral.collegeboard.com). Much of what I'm describing in this book represents the information the College Board has released on their site. Other information in this book comes from materials distributed to teachers at College Board–sponsored workshops as good a feel for the AP Physics 1 exam as it will be possible to get in the first few years that the exam is offered.

As you may have already noticed, the College Board has released only a limited number of sample questions for Physics 1. (That's in contrast to AP Physics C, which comes with authentic released exams going back more than four decades.) However, the content and scope of the questions that could appear on the Physics 1 Exam have been carefully defined in the *Curriculum Framework*, a 150-page tome less readable than *Finnegan's Wake*.[3] For now, the best anyone can do is read the *Curriculum Framework* carefully, parsing out which topics will be covered, in what depth and detail they'll be covered, and what types of test items will be used to cover those topics.

The good news is that this book has done all that dirty work for you. You don't need to bother reading the *Curriculum Framework*. If instead you read the chapters in this book and work through the practice items, you will have as good a feel for the AP Physics 1 Exam as it will be possible to get in this first year that the exam is offered.

What Is the Format of the AP Physics 1 Exam?

The following table summarizes the format of the AP Physics 1 Exam.

Table 1.1 AP Physics 1 Exam Structure

	AP PHYSICS 1		
SECTION	NUMBER OF QUESTIONS	TIME LIMIT	PERCENT OF SCORE
I. Multiple Choice	50	1 hour and 30 minutes	50%
II. Free Response	5	1 hour and 30 minutes	50%

What Types of Questions Are Asked on the Exam?

The multiple-choice questions all have four choices. Most are traditional multiple-choice questions, the kind you are already familiar with. But a five-question subsection of the multiple-choice portion is designated as "multiple correct" questions: you will be asked to

[3]But likely with more literary value.

choose *two* of the answers as correct. On these questions, you must mark both of the correct choices in order to earn credit.

The free-response section includes two short problems similar in style to end-of-chapter textbook problems; they include open-ended problem solving, as well as "justify-your-answer," verbal-response items. Another short problem requires a written response in paragraph form. One of the longer free-response questions is posed in a laboratory setting, asking for descriptions of experiments and analyses of results. The other long question is called the "qualitative-quantitative translation," which asks you to solve a problem numerically or symbolically and then explain in words how you got to your solution and what the solution means.

More details about these kinds of questions and how to deal with them can be found in Chapter 7 ("Strategies to Approach the Questions on the Exam") of this book.

Who Writes the AP Physics Exam?

Development of each AP Exam is a multiyear effort that involves many folks. At the heart of the effort is the AP Physics Development Committee, a group of college and high school physics teachers who are typically asked to serve for three years. The committee and other physics teachers create a large pool of multiple-choice questions. With the help of the testing experts at Educational Testing Service (ETS), these questions are then pretested with college students for accuracy, appropriateness, clarity, and assurance that there is no ambiguity in the choices. The results of this pretesting allow each question to be categorized by degree of difficulty. After several more months of development and refinement, Section I of the exam is ready to be administered.

The free-response questions that make up Section II go through a similar process of creation, modification, pretesting, and final refinement so that the questions cover the necessary areas of material and are at appropriate levels of difficulty and clarity. The committee includes the chief reader of the exams, who ensures that the proposed free-response problems can be graded consistently, fairly, and rapidly. The ETS specialist works with the committee to ensure that topic coverage and the scope of the exam are appropriate; the specialist makes sure that the exam tests what it's supposed to test.

At the conclusion of each AP reading and scoring of exams, the exam itself and the results are thoroughly evaluated by the committee and by ETS. In this way, the College Board can use the results to make suggestions for course development in high schools and to plan future exams.

What Topics Appear on the Exam?

The *Curriculum Framework* says nothing about the units or topics typically taught in an introductory physics class. Instead, the *Framework* is organized around six "Big Ideas" of physics that are each exemplified in numerous topics. So it's not possible to say exactly what topics are covered and to what extent. However, a careful reading of the *Curriculum Framework* can give a hint.

The eight topics listed here represent my own categorization of the material covered in AP Physics 1:

- Kinematics
- Forces
- Gravitation

- Impulse-Momentum
- Energy
- Rotation: Motion, Torque, Angular Momentum
- Electricity: Charge and Circuits
- Mechanical Waves, Sound, Simple Harmonic Motion

Do I Get to Use a Calculator? An Equation Sheet?

Well, yes. But please don't expect these things to help you much. The course is not about numbers and equations. If you come into the exam thinking you'll find the right equation on the equation sheet and then solve that equation with a calculator, you're going to be blown out of the water. In fact, I wish the College Board had decided *not* to allow calculators and equation sheets. They give a false sense of what kinds of questions will be asked on the exam and of how to prepare for them. (See Chapter 7 for more information about the types of questions you will encounter.)

Suffice it to say that you don't need the equation sheet because by test day, you will already know and understand the important relationships between quantities that underlie the physics questions that will be asked. And if you don't know the correct relationship, I don't advise picking through the dense and incomprehensible equation sheet; you're more likely to waste time than to find something useful there.

Regarding the calculator, you probably shouldn't use it more than a few times on the entire exam. Most problems won't involve calculation at all, but rather reasoning with equations and facts. Many of the problems that at first glance look like calculations can be solved more quickly and easily with semiquantitative reasoning.[4] The few problems that do require calculation will usually involve straightforward arithmetic (e.g., the mass of the cart will be 1 kg or 0.5 kg, not 0.448 kg).

How Is My Multiple-Choice Section Scored?

The multiple-choice section of the AP physics exam is worth half of the final score. Your answer sheet is run through the computer, which adds up your correct responses. The number of correct responses is your raw score on the multiple-choice section. No partial credit is awarded, even for the "multiple correct" items—either you choose both of the right answers, or you don't.

If I Don't Know the Answer, Should I Guess?

Yes. There is no penalty for guessing.

Who Grades My Free-Response Questions?

Every June, a group of physics teachers gathers for a week to assign grades to test takers' hard work. Each of these readers spends a day or so getting trained on only one question. Because each reader becomes an expert on that question, and because

---

[4]By *semiquantitative reasoning* I mean something like, "If I double the net force with the same mass, I also double acceleration by $F_{net} = ma$. So the new acceleration is twice the old acceleration of 1.2 m/s per second, so the answer is 2.4 m/s per second."

each exam book is anonymous, this process provides for consistent and unbiased scoring of that question.

During a typical day of grading, a random sample of each reader's scores is selected and cross-checked by experienced "table leaders" to ensure that consistency is maintained throughout the day and the week. Each reader's scores on a given question are also statistically analyzed to make sure scores are not given that are significantly higher or lower than the mean scores given by other readers of that question.

Will My Exam Remain Anonymous?

You can be absolutely sure that your exam will remain anonymous. Even if your high school teacher happens to randomly read your booklet, there is virtually no way he or she will know that exam is yours.[5] To the reader, each student is a number, and to the computer, each student is a bar code.

What about that permission box on the back? The College Board uses some exams to help train high school teachers so that they can help the next generation of physics students to avoid common mistakes. If you check this box, you simply give permission to use your exam in this way. Even if you give permission, your anonymity is maintained.

How Is My Final Grade Determined and What Does It Mean?

Each section counts for 50 percent of the exam. The total composite score is thus a weighted sum of the multiple-choice and free-response sections. In the end, when all of the numbers have been crunched, the chief faculty consultant converts the range of composite scores to the five-point scale of the AP grades.

This conversion is not a true curve; it's not that there's some target percentage of 5s to give out. This means you're not competing against other test takers. Rather, the five-point scale is adjusted each year to reflect the same standards as in previous years. The goal is that students who earn 5s this year are just as strong as those who earned 5s in 2000 or 2005.

Historically, it has taken about 60 to 65 percent of the available points on the AP Physics exams to earn a 5; it has taken about 50 percent of the points to earn a 4. I've used similar percentages in the tables at the end of the practice exams in this book to give you a rough example of a conversion. When you complete the practice exams, you can use this to give yourself a hypothetical grade.

Remember, AP Physics 1, Algebra-Based, is not just new, it's revolutionary. Everyone who is not associated with ETS is flying blind in figuring out what exactly is necessary to earn a score of 5, 4, or 3. All we can say for certain is that you are *not* expected to get classroom-style scores of 90 percent for an A. The exam is intended to differentiate between levels of students, and the exam tests far more than pure recall, so 60 percent is a strong score, not a weak score.

You should receive your AP grade in early July.

---

[5] Well, unless you write something like, "Hi, please kick Mr. Kirby in the butt for me. Thank you! Sincerely, George."

How Do I Register and How Much Does It Cost?

If you are enrolled in AP Physics at your high school, your teacher will provide all of these details, but a quick summary here can't hurt. After all, you do not have to enroll in the AP course to register for and complete the AP Exam. When in doubt regarding registration procedures, the best source of information is the College Board's website (https://www.collegeboard.org).

In 2013, the fee for taking the exam was $89. Students who demonstrate financial need may receive a reduction to offset the cost of testing. The fee and the reduction usually change from year to year. You can find out more about the exam fee, fee reductions, and subsidies from the coordinator of your AP program or by checking information on the College Board's website.

I know that seems like a lot of money for a test. But you should think of this $89 as the biggest bargain you'll ever find. Why? Most colleges will give you a few credit hours for a good score. Do you think you can find a college that offers those credit hours for less than $89? A credit hour usually costs hundreds of dollars. You're probably saving thousands of dollars by earning credits via AP.

There are several optional fees charged if you want your scores rushed to you or if you wish to receive multiple-grade reports. Don't worry about doing that unless your college demands it. (What? Do you think your scores are going to change if you don't find them out right away?)

The coordinator of the AP program at your school will inform you where and when you will take the exam. If you live in a small community, your exam may not be administered at your school, so be sure to get this information.

What If My School Doesn't Offer AP Physics at All? How Can I Take the Exam?

Any high school student is allowed to register for the exam, not just those who are taking an officially designated AP Physics course.

If your school doesn't offer any of the four AP Physics courses, then you should look at the content outlines and talk to your teacher. Chances are, you will want to take the AP Physics 1, Algebra-Based Exam, and chances are that you will have to do a good bit of independent work to delve deeper than your class discussed and practice the verbal responses necessary on this new exam. If you are a diligent student in a rigorous course, you will probably be able to do fine.

Your counseling office will be able to give you information about how to sign up for and where to take the test.

What Should I Bring to the Exam?

On exam day, I suggest bringing the following items:

- Several pencils and an eraser that doesn't leave smudges
- Black or blue pens for the free-response section[6]
- A ruler or straightedge

---

[6]Yes, I said *pens*. Your rule of thumb should be to do graphs in pencil and everything else in pen. If you screw up, cross out your work and start over. Then if you change your mind about what you crossed out, just circle it and say, "Hey, reader, please grade this! I didn't mean to cross it out!"

- A watch so that you can monitor your time (You never know if the exam room will have a clock on the wall. Make sure you turn off all beeps and alarms.)
- Your school code
- Your photo identification and social security number
- Tissues
- Okay, fine, a calculator, if it makes you happy (Don't you dare use it more than a few times.)
- Your quiet confidence that you are prepared (Please don't study the morning before the exam—that won't do you any good. Stop the studying the night before, and relax. Good luck.)

CHAPTER 2

Understanding the Exam: The AP Physics 1 Revolution

IN THIS CHAPTER

Summary: The AP Physics 1, Algebra-Based Exam requires less calculation and more written explanations of physics than any previous standardized physics exam. This chapter provides a deeper analysis of the new AP Physics 1, Algebra-Based Exam, explaining what the test is like and how it is different from traditional physics tests.

Key Ideas

✪ The AP Physics 1, Algebra-Based Exam is less focused on getting the "right" numerical answer to a problem and more focused on explaining and applying the concepts of physics.

✪ The AP Physics 1, Algebra-Based Exam is no walk in the park. Although it covers fewer topics, has fewer questions, and contains less math than the old AP Physics B Exam, it requires a deeper understanding of physics. Most students—especially those unprepared for this new type of physics exam—will probably find it more difficult.

What Happened to the AP Physics Test?

The new AP Physics 1, Algebra-Based Exam is not like anything your father or even your older sister took. The AP Physics 1, Algebra-Based curriculum has undergone a radical transformation that has eliminated the advanced mathematics but added a profound understanding of the science of physics.

A Little History

In the 1970s and 1980s, a typical physics professor gave lectures heavily focused on mathematics to first-year students and then administered exams that demanded clever algebraic manipulation. The AP Physics B exams of that era reflected the mathematical nature of college physics.

In the 1990s, a new generation of physics professors promoted a different kind of physics teaching that was just as rigorous but that demanded more explanations and less algebraic skill. The meaning of the math became more important than the math itself. AP exams began to include more questions asking for descriptions of experimental techniques, for justification of numerical answers, and for explanations in words.

When the College Board received a grant to redesign their science courses in the early 2000s, the curriculum design committee decided to move even farther away from calculation and more toward verbal explanation. They minimized the number of topics on the newest AP physics exams, and they decreased the number of questions on the exams. That way, they reasoned, the exam could demand more writing and more detailed explanations in the responses.

Goal of the AP Physics 1 Revolution: The *Best* College Physics Course, Not Just the "Typical" College Physics Course

The fundamental purpose of the College Board's AP program has always been to give advanced high school students access to college-level coursework. Historically, the goal was to create exams that mimic the content and level of the typical, average courses at an American university.

However, in redesigning their science courses, the development committees aimed higher—the stated goal now is for the AP exams to reflect best practices as well as the content and difficulty level of only the best college courses. In general, the "best" college classes include lots of demonstrations, laboratory work, and descriptive as well as calculational physics.

What Is AP Physics 1? Eleven Things You Should Know About the Course and Exam

You'll do better on a test if you understand the test and what's being tested. In this section you'll learn key facts about the AP Physics 1 Exam—facts that will help you know what to expect and, as a result, better prepare for the test.

1. AP Physics 1 Is Not a Broad Course

Whereas the old AP Physics B course included between five and eight major topic areas, AP Physics 1 is limited to just three:

- Mechanics
- Electricity
- Waves

Of these, mechanics will dominate the exam. Subtopics of mechanics primarily include motion, force, momentum, energy, and rotation. The study of electricity is limited to circuits and the force between charged particles. As an approximation, AP Physics 1 contains only a bit more than half as much material as the old AP Physics B course did. And that makes sense—AP Physics 1 is designed to replicate the first semester of an algebra-based freshman college course. AP Physics 2 covers the second-semester material.

2. AP Physics 1 Is Designed to Be a First-Time Introduction to Physics

The current AP Physics C exams, and the old AP Physics B Exam, were developed with the understanding that students would already have taken a high-school-level introduction to physics. But many high school students want to, and are ready to, dive right into algebra-based physics at the college level. AP Physics 1 has been written to set up these students for success. Thus, there is a reduction in the amount of material covered. Even if you've never seen physics before, you will have the time in AP Physics 1 to develop both your content knowledge and your physics reasoning skills enough to perform well on the exam.

3. AP Physics 1 Is Not a Math Course

There are only three high-school-level mathematical skills you need in order to understand AP Physics 1 material:

- You must have thorough facility with algebraic equations in a single variable.
- You must be able to calculate, and to understand the meaning of, the slope and area of a graph.[1]
- You must be able to use the definitions of the basic trig functions sine, cosine, and tangent.[2]

That's it. You covered these things in your Algebra 1 and geometry courses. You don't need matrices, factoring of polynomials, the quadratic formula, trigonometric identities, conic sections, or whatever else you are studying in Algebra II or precalculus.

4. AP Physics 1 Is Not About Numbers

Yes, you must use numbers occasionally. Yet you must understand that the number you get in answer to a question is always subordinate to what that number represents.

Many misconceptions about physics start in math class. There, your teacher shows you how to do a type of problem, and then you do several variations of that same problem for homework. The answer to one of these problems might be 30,000,000, and another 16.5. It doesn't matter . . . in fact, the book (or your teacher) probably made up random numbers to go into the problem to begin with. The "problem" consists of manipulating these random numbers a certain way to get a certain answer.

In physics, though, *every number has meaning*. Your answer will not be 30,000,000; the answer may be 30,000,000 joules, or 30,000,000 seconds, but not just 30,000,000. If you don't see the difference, you're missing the fundamental point of physics.

We use numbers to represent *real* goings-on in nature. The amount 30,000,000 joules (or 30 megajoules) is an energy; it could be the kinetic energy of an antitank weapon or the gravitational energy of an aircraft carrier raised up a foot or so.[3] Thirty million seconds is a time, not a few hours or a few centuries, but about one year. These two "30,000,000"

---

[1]Note that the calculus extensions of these concepts as "derivatives" and "integrals" are utterly irrelevant to and useless in AP Physics 1.

[2]This knowledge is often expressed as "SOHCAHTOA": In a right triangle, the **S**ine is **O**pposite over **H**ypotenuse. . . .

[3]Interestingly, when you use this much electrical energy in your house, it would cost you in the neighborhood of $1.

responses mean entirely different things. If you simply give a number as an answer, you're doing a math problem. It is only when you can explain the meaning of a result that you may truly claim to understand physics.

5. AP Physics 1 Requires Quantitative and Semiquantitative Reasoning

You'll only occasionally be asked to make numerical calculations. But you'll often be required to use mathematical *reasoning*.

"Quantitative" reasoning means not only performing direct calculations, but also explaining why those calculations come out the way they do. You'll be asked to explain whether quantities increase or decrease just by looking at the relevant equation, and without necessarily performing the calculations. You'll need to recognize that a problem is solvable when one equation with a single variable can be written, or when two equations with two variables can be written, or even when three equations with three variables can be written. You'll not be asked to solve even mildly complicated multivariable problems, but you must recognize when and explain why they are or aren't solvable.

"Semiquantitative" reasoning involves anticipating how the structure of an equation will affect the result of a calculation, even when no values in the equation are known. Increasing a variable in the numerator of an equation increases the quantity being calculated; increasing a variable in the denominator decreases the quantity being calculated. Doubling a term in the numerator also doubles the entire quantity, except if that term is squared (in which case the quantity is quadrupled) or square-rooted (in which case the quantity is multiplied by 1.4).

For example, the question "Calculate the acceleration of the 250-g cart" will be legitimate on the AP Physics 1 Exam. But more often the question will be rephrased to get at the heart of your ability to reason about physics, not just at your math skills. For example:

- Rank the accelerations of these carts from greatest to least; justify your ranking.
- Can the acceleration of the 250-g cart be calculated from the given information? If not, what other information is required?
- Explain how you would use a graph to determine the acceleration of the 250-g cart.
- How would the acceleration of a 500-g cart compare to the acceleration of the 250-g cart?
- Is the acceleration of the cart greater than, less than, or equal to g?

6. AP Physics 1 Requires Familiarity with Lab Work

Every physics problem that asks for calculation, quantitative reasoning, or semiquantitative reasoning is, in truth, asking for an experimental prediction.

When doing calculations, what differentiates physics from math class? In physics, every calculation can, in principle, be verified by an experimental measurement. Merely using an equation to calculate that a cart's acceleration is 2.8 m/s per second is a math problem, one that you might see in an Algebra 1 class. It's only a physics problem because you can, in fact, go to a cabinet and pull out a 250-g cart and a motion detector that will measure the cart's acceleration. If you set up the situation that was described in the problem, you'd better get an acceleration of 2.8 m/s per second; otherwise, either the calculational approach was incorrect (e.g., the equation you used might not apply to this situation), or the experiment was set up inappropriately (e.g., the problem assumed a level surface, but your track was slanted).

What's particularly nice about AP Physics 1 is that almost every problem posed in this course can be set up for experimental measurement within the realm of most students' experience. Carts can be set up to collide in your classroom. It's straightforward to take smartphone video of cars on a freeway or of a roller coaster. Computerized data collection—generally using equipment from PASCO or Vernier—should be part of your classroom experience, so that you're familiar with using force probes, motion detectors, photogates, and so on.[4]

Any physical situation can spawn a laboratory-based experimental question. Be prepared to describe experiments and to analyze data produced by experiments:

- Describe an experiment which uses commonly available laboratory equipment to measure the acceleration of the 200-g cart.
- In the laboratory, this table of the cart's speed as a function of time was produced. Use the data to determine the car's acceleration.
- The acceleration of the cart is calculated to be 2.8 m/s per second, but in the laboratory, a student measures the cart to have an acceleration of 4.1 m/s per second. Which of the following might explain the discrepancy between theory and experiment?

7. AP Physics 1 Does Not Require That You Perform a Specific Set of Programmed Laboratory Exercises

In AP Biology, students are expected to be familiar with, and to have actually done, a set of experiments. Biology exam questions will refer to these common experiments, expecting prior knowledge to carry students. This approach is completely different from that in AP Physics.

AP Physics certainly requires experimental skills, as described above. Teachers are required by the Course Audit to spend at least 25 percent of class time doing live, hands-on laboratory exercises. Yet, the actual nature of those exercises is left to the teacher. Since virtually every possible AP Physics 1 question can be set up as an experiment, there are limitless possibilities for lab work. Creativity in lab work is prized on the AP Physics exam.

It's critical that you don't think of the "lab" as a place where you follow the steps in a procedure to produce a canned result that matches your teacher's expectation. Rather, think of the lab as a place to play, a place to re-create the calculational problems you've been solving in class. Lab is a place where you test the equations and concepts to see if they work.

> That's silly. Of course these equations and concepts work. Do you really expect me, a high school student, to disprove the conservation of momentum in a collision? Really!

Yeah, no one expects you to win a Nobel Prize in your AP Physics 1 laboratory. What is expected is that you go beyond stating "facts" of physics as gospel. A physicist always asks, "What's the evidence?" "How do we know that?" The AP Exam expects you to be able to use conservation of momentum to calculate the speed of a cart after collision, sure; but it also expects you to explain why conservation of momentum is valid in this situation and, this is important, to explain *what evidence exists that conservation of momentum is a valid principle in the first place.*

The "evidence" for each physics fact comes from an experiment. You should be able to articulate how an experiment could be designed to test the validity of any fact; you should be able to use equipment creatively to make a measurement of any quantity that might show up in a calculational problem.

---

[4]If your classroom does *not* have at least one set of PASCO or Vernier probes, your school probably doesn't meet the requirements of the AP Course Audit. Access to some of the same type of laboratory equipment that is available in most colleges is a prerequisite for the College Board, allowing a school to label its course as "AP." See the "AP Central" portion of the College Board's official website for details about the course audit.

8. AP Physics 1 Requires Writing

The free-response questions on the AP Physics 1 Exam will be more similar to those on the AP Biology or AP Economics exams than to those on AP Calculus or AP Physics C exams. Do not expect to answer exclusively in mathematical symbols and numbers. Whereas AP Physics B questions rarely required more than a couple of sentences at a time, the new exam will ask for short answers, descriptions, and explanations "without equations or calculations."

This doesn't mean you need to develop your storytelling skills. The writing required is always straightforward and to the point. A perfect response on the AP Physics 1 Exam might draw all sorts of complaints from your English teacher. While you need to use (reasonably) grammatically correct sentences, your language and vocabulary should be simple, not flowery. Your sentence structure doesn't need to be varied and interesting. You don't need to grab your reader's attention or to segue appropriately between paragraphs and ideas.

Just write, without worrying about how your writing sounds to a professional. Imagine that the person reading your writing is a student at the same level of physics as you. Don't explain your answers the way you think a college professor would; explain your answers the way you wish your teacher would explain them—simply and clearly, but completely. Practice this sort of writing on your problem sets.

9. AP Physics 1 Requires "Multiple Representations" of Physics Concepts

This means you should be comfortable explaining physics with words, equations, diagrams, and numbers. When you solve problems in your physics class, practice using these elements in every solution—even if your teacher doesn't explicitly require them. In my own class, if a student doesn't use at least three of these four elements in response to a homework question, the student usually loses significant credit.

Don't be shy about drawing diagrams. Try doing your homework on graph paper or unlined paper, rather than on standard notebook paper. Think of the paper as a blank canvas that needs to be filled in with your understanding of the solution to a problem. Lined notebook paper is far too restricting. It implies that you should be writing rows of words, with maybe an equation.[5] Words can be written on a diagram, possibly with arrows to point out the important parts. Graphs and pictures can be drawn anywhere. Equations don't have to be placed one after the other in columns. A series of equations, however you present them, should always include some words describing the purpose of the equations in the problem's solution.

Then on the exam, you'll be well practiced in interpreting every possible representation of a physics explanation. Don't worry, you'll be asked for such interpretations on the multiple-choice section, and you'll be asked to use multiple representations of concepts on the free-response questions.

10. The AP Physics 1 Exam Is Designed to Give You the Time You Need to Answer the Questions Posed

As the new exams were in the development process, it became clear that the heightened demands for writing, for experimental interpretation and description, and for multiple representations of physics concepts would require a lot of time and thought. The committees in charge of creating AP Physics 1 were in agreement that students must be given the time necessary to respond in a complete way.

---

[5] And that equation's fraction bar, if there is one, screws up the look of the words.

Even students with strong physics abilities are often pressed for time on the AP Physics C or the old AP Physics B exams. The rule of thumb there was to spend about one minute per point: one minute on a multiple-choice question, 15 minutes on a 15-point free-response question, and so on. Therefore, students were advised to work quickly, eschewing in-depth thought for quick solution methods.

But AP Physics 1 is about half as long. The rule of thumb will be to spend in the neighborhood of *two* minutes per point. You'll get 50 multiple-choice questions to be completed in 90 minutes. Knock a few off quickly, and you can really think carefully about some while easily maintaining a just-under-two-minute-per-problem pace. The free-response section will have two long 12-point questions and three short 7-point questions and also will be 90 minutes in length. Yeah, you'll need to work steadily without dawdling, otherwise you might run out of time even on this exam. But you'll have time to think before you write; then you'll have time to write everything you need to communicate your answer.

Practice this new sort of time management on your homework for physics class. Instead of doing a zillion homework problems as quickly as you can, try picking one or two for a full-on, long-form treatment. Answer using multiple representations. Explain the answer, how you got it, how it would change if the problem inputs changed, and why you used the fundamental approach you used. Make the solution so complete that it could serve as the basis for a set of PowerPoint slides that could present your solution to a classmate. And do it all in less than 30 minutes. When you can do that, you're ready for the AP Physics 1 Exam.

11. AP Physics 1 Is a Difficult, High-Level College Course

Some may get the impression that because calculation is minimized and so few topics are covered, this course will be a "piece of cake." Wrong. My own impression, and the impression of all experts who have looked carefully at the types of questions asked on the new exam, is that AP Physics 1 is substantially more difficult than AP Physics B used to be.

People have the false impression that "lots of writing" means "easy to get some credit, because I can use lots of big words full of sound and fury." You will find that the AP readers are as adept at recognizing baloney as they are at recognizing good physics. You will be less likely to find points awarded for attempting to use a correct equation. While partial credit on the free response will still be copiously available, that credit will generally require a good, if incomplete, understanding of physics, and will not be attainable via guesswork.

Expect wailing and gnashing of teeth once scores come out after that first exam. Not by you, of course, because you've read this book; you know what to expect; and you know that a true understanding of physics requires that you be able to solve problems, explain how you solved them, explain what concepts you used to solve them, explain why those concepts apply, and explain how you could experimentally demonstrate your solution. Meanwhile, those with gnashed teeth remain stuck with the idea that just producing an answer is enough.

CHAPTER 3

How to Use Your Time

IN THIS CHAPTER

Summary: You'll need to set up a study plan that's personalized to fit your needs and the amount of time you have to prep for the test. This chapter provides information and advice to get you started and outlines what you can do if you have a full year, a semester, or only a few weeks left until test day.

Key Ideas

✪ It's not possible to "game" the test—you have to really know your physics. Last-minute cramming won't work since the exam tests your skills and understandings, not your ability to remember facts and formulas.

✪ The most important part of your test prep plan is your AP Physics 1, Algebra-Based course.

✪ Personalize your study plan. Focus your test prep on the topics and types of questions that you find the most troublesome.

✪ Essential elements of any test prep plan are (1) familiarizing yourself with the test, (2) learning the best strategies to use in approaching each of the question types, and (3) taking practice tests.

Personalizing Your Study Plan

First, it's important for you to know that the AP Exam is an authentic physics test. What this means is that it's not possible to "game" this test—in order to do well, *you must know your physics*. It's not possible to slack off and then cram for the test and expect to do well.

The most important part of your study plan is your AP Physics 1 class, which is in fact designed to teach the knowledge and skills required on the exam. Diligent attention to all your lectures, demonstrations, and assignments will save you preparation time in the long run. (See Chapter 6 for strategies to get the most out of your class.)

Your study plan should be personalized based on your needs. Use the diagnostic tools in Step 2 to identify your weaknesses and then build a plan that focuses on these. If you're comfortable with kinematics and projectile problems, why would you spend any time on these? On the other hand, if you're worried about, say, collisions, then spend a couple of evenings reviewing and practicing how to deal with them. Focusing on weaknesses, rather than starting at the beginning and trying to review everything, will allow you to use the time you have to produce the maximum benefit.

Every reader of this book will have a different study plan. You can't follow some one-size-fits-all timetable, and you shouldn't start at the beginning of your course and try to review absolutely everything—you won't be able to do it. Your study plan depends not only on the topics you most need to review, but also on the types of exam questions you find most difficult, and the amount of time you have to study. Develop a realistic plan, and stick to it.

Regardless of when you start to prepare or how much content you want to review, your study plan should include these essential elements:

- Familiarize yourself with the test (Chapters 1 and 2)
- Learn the best strategies to approach each type of test question (Chapters 7 through 9)
- Take complete practice tests (Step 5).

Plan A: You Have a Full School Year to Prepare

If you're opening this book at the beginning of the school year, you're off to a good start. Here's what you can do:

First Semester
- Read Chapter 5 on how to get the most out of your AP Physics 1 class.
- Begin to familiarize yourself with the test (Chapters 1 and 2).
- Start practicing the strategies to approach the different types of questions found on the test (Chapters 7 through 9).
- You can work through the review chapters in this book (Chapters 10 to 18) as you study the same topics in your AP course. This will help you by providing a different perspective on the key content and ensuring you really understand the physics you need to know. Practice using the strategies presented in Chapters 7 through 9 to approach the test-like, free-response questions at the end of each review chapter.

Second Semester
- Keep working through the review chapters as you progress through the physics course.
- About three months before the AP Exam, use the diagnostic tools in Chapters 4 and 5 to assess your weaknesses. Try to identify both the content areas and types of questions you have the most difficulty with. Then focus your test prep review on the weaknesses you identify.

Six Weeks Before the Test
- Continue to focus on the areas of weakness that you identified based on the self-tests in Chapters 4 and 5.

- Review the strategies in Chapters 7 through 9 to make sure you approach the questions in ways that will help you get your best score.
- Be sure to take the practice tests (Step 5) a couple of weeks before the real test. These tests closely resemble the actual AP Physics 1 Exam. They will help you learn to pace yourself and allow you to experience what the test is really like. Then focus any final content review on the areas that proved troublesome on the practice tests. Also allocate some time to practice answering the question types that gave you the biggest problems.
- Be confident. You've worked all year, and you're really set to do your best.

Plan B: You Have One Semester to Prepare

Most students begin a test prep plan in the second semester. You should still have time to use this book to familiarize yourself with the test, learn the best strategies to approach each type of question, review the topics you find most troublesome, and take practice tests. Here's what you can do:

Second Semester
- Read Chapter 5 on how to get the most out of your AP Physics 1 class.
- Familiarize yourself with the test (Chapters 1 and 2).
- Practice the strategies to approach the different types of questions found on the test (Chapters 7 to 9).
- Two or three months before the AP Exam, use the diagnostic tools in Chapters 4 and 5 to assess your weaknesses. Try to identify both the content areas and types of questions you have the most difficulty with. Then focus your test prep review on the topics you identify.

Six Weeks Before the Test
- Continue to focus on the areas of weakness that you identified based on the self-tests in Chapters 4 and 5.
- Review the strategies in Chapters 7 through 9 to make sure you approach the questions in ways that will help you get your best score.
- Be sure to take the practice tests (Step 5) a couple of weeks before the real test. They will help you learn to pace yourself and allow you to experience what the test is really like. Then focus any final content review on the areas in which you didn't do as well. Also practice the question types that proved problematic.
- Be confident. You know the test and the best strategies for the different question types. In addition, you've reviewed the areas in which you felt weakest, and you've taken a practice test. You're set to get a good score.

Plan C: You Have Six Weeks to Prepare

Six weeks should be plenty of time to prepare for the AP Exam. You've been working diligently in your class all year, learning the topics as they've been presented. These last weeks should be spent putting it all together. Focus on the topics you feel you most need to review and on problems that cover multiple concepts in one question. Use your test prep time not to try to cram for the test (it won't work), but to familiarize yourself with the exam and learn the best strategies to use to approach the different question types. Most important, take the practice tests in Step 5 of this book. Here's what you can do:

Six Weeks Before the Test
- Familiarize yourself with the test (Chapters 1 and 2).
- Practice the strategies to approach the different types of questions found on the test (Chapters 7 to 9).
- Use the diagnostic tools in Chapters 4 and 5 to assess your weaknesses. Try to identify both the content areas and types of questions you have the most difficulty with. Then pick only a few areas on which to focus your test prep review and learn them well. Don't try to cover everything.
- Be sure to take the practice tests (Step 5) at least a week before the real test. They will help you learn to pace yourself and give you a trial run so you can experience what the test is really like. Then focus any final content review on the areas on which you didn't do as well. Also focus on the question types that gave you the biggest problems.
- Be confident. You know the test and the best strategies for the different question types. In addition, you've reviewed the areas in which you felt weakest and you've taken a practice test. If you've also used your AP course to really learn physics, you're set to get a good score.

STEP 2

Determine Your Test Readiness

CHAPTER 4   Test Yourself: AP Physics 1
             Fundamentals

CHAPTER 5   Test Yourself: AP Physics 1
             Question Types

CHAPTER 4

Test Yourself: AP Physics 1 Fundamentals

IN THIS CHAPTER

Summary: This chapter contains a short test designed to help you determine your strengths and weaknesses regarding the content and skills tested on the AP Physics 1, Algebra-Based Exam.

Key Ideas

✪ Find out what you know—and don't know—about mechanics, electricity, and waves. This will tell you how well you're prepared for the subjects tested on the AP Physics 1, Algebra-Based Exam.

✪ From this self-assessment you can identify strengths and weaknesses and develop a personalized test-prep plan (see Chapter 3).

Self-Assessment: AP Physics 1 Fundamentals

Note that the questions in this self-assessment are *not* written in the style of the actual questions on the AP Physics 1, Algebra-Based Exam. The questions are designed to quickly determine your strengths and weaknesses, not to mimic actual test questions. In the next chapter you'll encounter the different question types found on the actual AP Physics 1 Exam.

Answer the questions below. The correct answers with explanations are found at the end of this chapter. From this self-assessment you should get a sense of what your weaknesses are and be able to prioritize what areas you need to give the most attention. This self-assessment should be the basis for a test-prep plan that you develop for yourself (see Chapter 3).

Mechanics

1. What is the mass of a block with weight 100 N?

2. Give the equations for two types of potential energy, identifying each.

3. When an object of mass m is on an incline of angle θ, one must break the weight of the object into components parallel and perpendicular to the incline.
   (a) What is the component of the weight parallel to the incline?
   (b) What is the component of the weight perpendicular to the incline?

4. Write two expressions for work, including the definition of work and the work-energy principle.

5. Name several things that can *never* go on a free-body diagram.

6. Write two expressions for impulse. What are the units of impulse?

7. In what kind of collision is momentum conserved? In what kind of collision is kinetic energy conserved?

8. What is the mass of a block with weight W?

9. A ball is thrown straight up. At the peak of its flight, what is the ball's acceleration? Be sure to give both magnitude and direction.

10. A mass experiences a force with components 30 N to the right, 40 N down. Explain how to determine the magnitude and direction (angle) of the force.

11. Write the definition of the coefficient of kinetic friction, μ_k. What are the units of μ_k?

12. How do you find acceleration from a velocity-time graph?

13. How do you find displacement from a velocity-time graph?

14. How do you find velocity from a position-time graph?

15. A cart on a straight track has a positive acceleration. Explain BRIEFLY how to determine whether the cart is speeding up, slowing down, or moving at constant speed.

16. Given the velocity of an object, how do you tell in which direction that object is moving?

17. When is the gravitational force on an object mg? When is the gravitational force Gm_1m_2/r^2?

18. What is the direction of the net force on an object that moves in a circle at constant speed?

19. Under what conditions is the equation $x - x_0 = v_0t + \frac{1}{2}at^2$ valid? Give a specific situation in which this equation might seem to be valid, but is *not*.

20. Under what conditions is angular momentum conserved?

21. What is rotational inertia, and how is it calculated?

Electricity

22. What is the equation for the force of one charge on another?

23. What property of series resistors is the same for each and equal to the total?

24. What property of series resistors is different for each, adding to the total?

25. What property of parallel resistors is the same for each and equal to the total?

26. What property of parallel resistors is different for each, adding to the total?

27. The brightness of a bulb depends on what physical quantity?

Waves

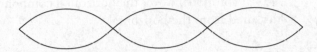

Questions 28 and 29 relate to the preceding diagram of a standing wave.

28. On the diagram, label one wavelength.

29. The diagram represents particle displacement for a longitudinal wave in a pipe. Is this a pipe closed at one end, or a pipe open at both ends?

30. How are period and frequency related?

31. The Doppler effect affects which property of a wave?

32. A guitar string is tightened. What variable in $v = \lambda f$ is *not* affected?

Solutions for the AP Physics 1 Fundamentals Self-Assessment

Mechanics

1. Weight is *mg*. So, mass is weight divided by *g*, which would be 100 N/(10 N/kg) = 10 kg.

2. PE = *mgh*, gravitational potential energy;

   PE = $\frac{1}{2}kx^2$, potential energy of a spring.

3. (a) It is *mg* sin θ is parallel to the incline.
   (b) It is *mg* cos θ is perpendicular to the incline.

4. The definition of work is work = force times parallel displacement.

   The work-energy principle states that $W_{NC} = (\Delta KE) + (\Delta PE)$

5. Only forces acting on an object and that have a single, specific source can go on free-body diagrams. Some of the things that cannot go on a free-body diagram but that students often put there by mistake include the following:

motion	mass	acceleration	*ma*
centripetal force	velocity	inertia	

6. Impulse is force times time interval, and also change in momentum. Impulse has units either of newton·seconds or kilogram·meters/second.

7. Momentum is conserved in *all* collisions. Kinetic energy is conserved only in elastic collisions.

8. Using the reasoning from question #1, if weight is *mg*, then *m = W/g*.

9. The acceleration of a projectile is *always g*; i.e., 10 m/s², downward. Even though the velocity is instantaneously zero, the velocity is still changing, so the acceleration is *not*

zero. (By the way, the answer "−10 m/s²" is wrong unless you have clearly and specifically defined the down direction as negative for this problem.)

10. The magnitude of the resultant force is found by placing the component vectors tip-to-tail. This gives a right triangle, so the magnitude is given by the Pythagorean theorem, 50 N. The angle of the resultant force is found by taking the inverse tangent of the vertical component over the horizontal component, $\tan^{-1}(40/30)$. This gives the angle measured from the horizontal.

11. $\mu = \dfrac{F_f}{F_n}$

    friction force divided by normal force. μ has no units.

12. Acceleration is the slope of a velocity-time graph.

13. Displacement is the area under a velocity-time graph (i.e., the area between the graph and the horizontal axis).

14. Velocity is the slope of a position-time graph. If the position-time graph is curved, then instantaneous velocity is the slope of the tangent line to the graph.

15. Because acceleration is not zero, the object *cannot* be moving with constant speed. If the signs of acceleration and velocity are the same (here, if velocity is positive), the object is speeding up. If the signs of acceleration and velocity are different (here, if velocity is negative), the object is slowing down.

16. An object *always* moves in the direction indicated by the velocity.

17. If the gravitational field g is known, mg gives the gravitational force. Newton's law of gravitation, Gm_1m_2/r^2, is valid everywhere in the universe.

18. An object in uniform circular motion experiences a *centripetal*, meaning "center-seeking," force. This force must be directed to the center of the circle.

19. This and all three kinematics equations are valid only when acceleration is constant. So, for example, this equation cannot be used to find the distance traveled by a mass attached to a spring. The spring force changes as the mass moves; thus, the acceleration of the mass is changing, and kinematics equations are not valid. (On a problem where kinematics equations aren't valid, conservation of energy usually is what you need.)

20. Angular momentum is conserved when no torques act on a system other than torques due to objects within the system itself.

21. Rotational inertia I is an object's innate resistance to a change in its angular velocity. For an object that can be treated as a point mass, $I = mr^2$, where m is the object's mass and r is the distance from the center of rotation. For multiple objects, add the rotational inertias of each object together to get the total.

Electricity

22. $F = k\dfrac{Q_1Q_2}{d^2}$

23. Current

24. Voltage

25. Voltage

26. Current

27. Power

Waves

28.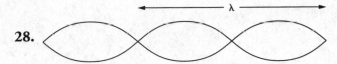

29. Open at both ends. The standing wave in such a pipe is symmetric, with a node at both ends (or an antinode at both ends).

30. They are reciprocals—the period is $1/f$.

31. Frequency and wavelength

32. Wavelength, λ. The length of the string doesn't change, and the wavelength is related to the length of the string. The speed v changes because the string is tightened, increasing the wave speed; the frequency f changes because a tighter string causes a higher pitch.

CHAPTER 5

Test Yourself: AP Physics 1 Question Types

IN THIS CHAPTER

Summary: In the previous chapter you tested yourself on physics fundamentals. In this chapter you will test yourself on the types of questions found in the AP Physics 1, Algebra-Based Exam. This preview of question types will provide insight into which types of questions will be the most problematic for you. Complete explanations for all the questions in this diagnostic test are included at the end of this chapter.

Key Ideas

✪ In addition to reviewing AP Physics 1, Algebra-Based content and skills, you will need to become familiar with the question types on the exam and assess which types of questions will be most difficult for you.

✪ This self-assessment, along with the one in Chapter 4, should be used to help you develop a personalized test-prep plan based on your needs (see Chapter 3).

Self-Assessment: Question Types

In the self-assessment that follows, you'll be introduced to the question types found on the AP Physics 1, Algebra-Based Exam. Sure, you've taken multiple-choice and essay tests before, but on the AP Physics 1 Exam, you'll find some multiple-choice and free-response questions unlike anything you've probably seen before. Your test score will improve if you not only review physics subject areas but also become familiar with the types of questions you will encounter and practice responding to these. Adjust your study plan (see Chapter 3) to focus on any weaknesses you identify as a result of this diagnostic test. In Chapters 7 through 9 you'll find proven strategies for attacking each of the question types on the AP Physics 1 Exam. Be sure to familiarize yourself with these, especially for those types of questions that proved difficult for you.

Descriptive Problems (Like Those You've Probably Seen Before)

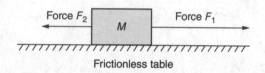

Frictionless table

1. In the preceding diagram, forces F_1 and F_2 are acting on box M which is on a frictionless table. F_1 has a greater magnitude than F_2. Of the following statements about the motion of box M, which is correct?

   (A) Box M is accelerating in the direction of F_1.
   (B) Box M is accelerating in the direction of F_2.
   (C) Box M is moving with a constant speed in the direction of F_1.
   (D) Box M is moving with a constant speed in the direction of F_2.

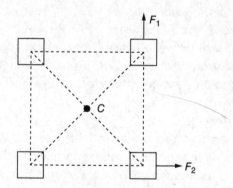

2. Four objects with mass m are rigidly connected and free to rotate in the plane of the page about the center point C. The objects experience two forces, F_1 and F_2, as shown in the preceding diagram. Which of the following statements correctly analyzes how F_1 and F_2 will affect the angular velocity of the objects?

   (A) Both forces apply torque in the same sense, so the angular velocity must increase.
   (B) If the forces F_1 and F_2 are equal in magnitude, the angular velocity will not change.
   (C) The angular velocity could decrease if the objects were initially rotating clockwise.
   (D) F_2 does not apply any torque to the objects, so only F_1 can cause a change in angular velocity.

Calculation Problems (Like Those You've Probably Seen Before)

3. A 2.0-kg ball is dropped such that its speed upon hitting the ground is 3.0 m/s. It rebounds, such that its speed immediately after collision is 2.0 m/s. What is the magnitude of the ball's change in momentum?

(A) 10 N·s
(B) 5 N·s
(C) 2 N·s
(D) 1 N·s

Handwritten: $M = 2kg$ $p = m(V_f - V_i)$
$V_i = 3.0 \, m/s$ $p = 2(1)$
$V_f = 2.0$

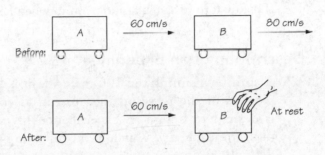

Before: A →60 cm/s B →80 cm/s

After: A →60 cm/s B At rest

4. Two carts, each of mass 0.5 kg, move to the <u>right</u> at different speeds as shown in the diagram. Next, a student stops cart B with his hand. By how much has the linear momentum of the two-cart system changed after the student stops cart B?

(A) 0.4 N·s
(B) 0.1 N·s
(C) 0.3 N·s
(D) 0.7 N·s

Handwritten: $P_I = 0.5(.6) + 0.5(.8)$
$P_I = 0.7$ $P = m(V_f - V_i)$
$P_F = 0.5(0.6)$
$P_F = 0.3$

Ranking Task

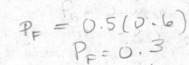

Initially at rest   Initially rotating clockwise by 5 rev/s   Initially rotating counter-clockwise by 5 rev/s
① ② ③
F   F   F

5. Three identical rods experience a single, identical force F, as shown in the diagrams. Each rod is <u>initially rotating</u> differently about its left edge, as described. Which of the following correctly ranks the magnitude of each rod's change in angular speed during the first 0.5 s that the force is applied?

(A) 2 = 3 > 1
(B) 1 = 2 = 3
(C) 1 > 2 > 3
(D) 3 > 2 > 1

Handwritten: $\tau = I\alpha$
$|\tau = I(\frac{\omega}{t})|$

Semiquantitative Reasoning

6. A diver leaps upward and forward off of a diving board. As the diver is in the air, he twists his body such that his rotational speed when he hits the water is four times his rotational speed when he left the diving board. When he hits the water, what is the diver's angular momentum about his center of mass?

$L = I \frac{\omega}{4}$

(A) Greater than his angular momentum when he left the diving board by a factor of four
(B) Greater than his angular momentum when he left the diving board, but not by a factor of four
(C) Less than his angular momentum when he left the diving board
(D) Equal to his angular momentum when he left the diving board

Description of an Experiment

7. A sprinter running the 100-meter dash is known to accelerate for the first few seconds of the race and then to run at constant speed the rest of the way. It is desired to design an experimental investigation to determine the sprinter's maximum speed v. Which of the following procedures could correctly make that determination?

(A) Place poles 90 m and 100 m from the race's start. Measure with a stopwatch the time t for the sprinter to travel between the poles. To find v, divide 10 m by t.
(B) Estimate that the sprinter accelerates for the first 2.5 s. Mark on the track the location of the sprinter after 2.5 s. Use a measuring tape to find the distance d the sprinter traveled in this time. Divide d by 2.5 s to get v.
(C) Measure with a stopwatch the time t for the sprinter to run the 100 m. To find v, divide 100 m by t.
(D) Measure with a stopwatch the time t for the sprinter to run the 100 m. Divide 100 m by t^2 to get the average acceleration a. Then since the sprinter starts from rest, v is given by $\sqrt{2(a)(100 \text{ m})}$.

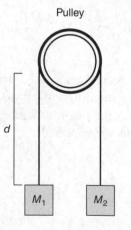

8. Two blocks of known masses M_1 and M_2, with $M_1 > M_2$, are connected by string over a freely rotating light pulley, as shown in the preceding diagram. A video camera records the motion of the blocks and pulley after the blocks are released from rest in the position shown. It is desired to use the video to measure the angular velocity of the pulley when block m_1 has fallen a known distance d. Which of the following approaches will best make this experimental determination?

(A) Treat the system as a single mass. The net force is the difference between the blocks' weights, $m_1 g - m_2 g$; Newton's second law gives an acceleration of $a = \left(\dfrac{m_1 - m_2}{m_1 + m_2} \right) g$. Use the kinematic equation $v_f^2 = v_0^2 + 2ad$ with $v_0 = 0$ to determine the final speed of the block; then the angular velocity is this speed divided by the radius of the pulley.

(B) Mark a spot on the edge of the pulley. Run the video until the blocks have gone the distance d. In that time, count the total revolutions that spot makes, including any partial revolutions measured using a protractor. Divide the total revolutions by the time the video ran to get the angular velocity.

(C) Pause the video when mass m_2 has just reached the distance d; note the location of a position on the rim of the pulley. Advance the video one frame. Use a protractor to measure the angle through which the noted location on the pulley moved in that one frame. Divide that angle by the camera's time between frames to get the angular velocity of the pulley.

(D) Make a graph of the position of mass m_1 as a function of time, determining the block's position by pausing the video after every frame. The slope of this graph is the pulley's angular velocity.

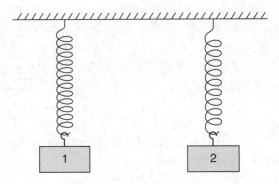

9. Two hanging blocks, each attached to a different spring, undergo oscillatory motion. It is desired to determine, without stopping the motion, which block experiences the greater maximum acceleration. Which of the following procedures would accomplish that determination?

(A) Place a motion detector underneath each block. On the velocity-time graphs output by the detector, look at the maximum vertical axis value, indicating the highest speed that block attained. Whichever block attains the higher speed has the larger acceleration.

(B) Place a motion detector underneath each block. On the velocity-time graphs output by the detector, look at the steepest portion of the graph. Whichever block makes the steeper maximum slope on the velocity-time graph has the greater maximum acceleration.

(C) Place a motion detector underneath each block. On the position-time graphs output by the detector, look at the maximum vertical axis value, indicating the amplitude of the motion. Whichever block oscillates with the larger amplitude has the larger acceleration.

(D) Place a motion detector underneath each block. On the position-time graphs output by the detector, look at the steepest portion of the graph. Whichever block makes the steeper maximum slope on the position-time graph has the greater maximum acceleration.

Analysis of an Experiment

10. In the laboratory, measured net torques τ are applied to an initially stationary pivoted bar. The resulting change in the bar's angular speed after 1 s is measured and recorded as $\Delta\omega$. Which of the following graphs will produce a slope equal to the bar's rotational inertia about the pivot point?

(A) τ versus $\Delta\omega$

(B) τ versus $\dfrac{1}{\Delta\omega}$

(C) τ versus $(\Delta\omega)^2$

(D) τ versus $\sqrt{\Delta\omega}$

Set of Questions Referring to the Same Stem

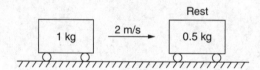

Questions 11 and 12: A 1-kg cart moves to the right at 2 m/s. This cart collides with a 0.5-kg cart that is initially at rest; the carts stick together after the collision. Friction on the surface is negligible.

11. What is the kinetic energy of the two-cart system after the collision?

(A) 1.3 J

(B) 0 J

(C) 0.7 J

(D) 2 J

12. If instead the carts collide elastically, which of the following is correct about the linear momentum and kinetic energy of the two-cart system after the collision compared to the collision in which the carts stuck together?

(A) The linear momentum and the kinetic energy will both be greater.

(B) The linear momentum will be greater, but the kinetic energy will be the same.

(C) The kinetic energy will be greater, but the linear momentum will be the same.

(D) The linear momentum and the kinetic energy will both be the same.

Multiple Correct Questions

Questions 13 through 15 are multiple correct: Mark the *two* correct answers.

13. (multiple correct) An astrophysicist is modeling the behavior of stars orbiting the center of the Milky Way galaxy. In his model, he must consider the effect of each of the four fundamental natural forces. Which of the following correctly indicates a negligible force with a correct explanation for neglecting that force? Choose two answers.

(A) The gravitational force is negligible, because the order of magnitude of the gravitational constant G (10^{-11} N·m²/kg²) is extraordinarily small compared to the order of magnitude of the Coulomb's law constant ($k = 10^9$ N·m²/C²).

(B) The electric force is negligible, because it cannot act beyond atomic ($\sim10^{-10}$ m) distances.

(C) The weak force is negligible because although it is responsible for a star's energy release via nuclear fusion, the nuclear nature of the weak force means that it is irrelevant for star-to-star interactions at a distance.

(D) The strong force is negligible because it cannot act beyond nuclear ($\sim10^{-15}$ m) distances.

14. (multiple correct) A car tire initially rotates clockwise with a rotational speed of 20 rad/s. The rotation gradually slows, such that 2 s later the tire rotates clockwise with a rotational speed of 10 rad/s. Considering clockwise as the positive direction, which of the following vectors is positive? Choose two answers.

(A) the tire's angular acceleration
(B) the tire's angular momentum
(C) the net torque on the tire
(D) the tire's angular velocity

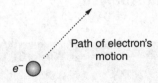

Path of electron's motion

e^-

15. (multiple correct) An electron in a vacuum chamber is moving at constant velocity in the direction shown in the preceding diagram. Which of the following force vectors F applied to the electron would increase the electron's kinetic energy? Choose two answers.

(A) F ↑

(B) F ←

(C) F ↗

(D) F ↙

Qualitative-Quantitative Translation

Note: The following question is part of a free-response question.

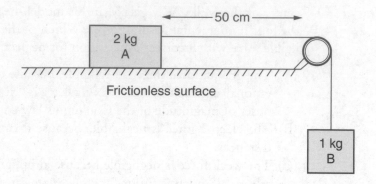

16. Two blocks are connected over a light, frictionless pulley, as shown. Block A of mass 2 kg is on a frictionless surface; block B of mass 1 kg hangs freely. The blocks are released from rest, with block A 50 cm from the end of the table.

    (a) Calculate the speed of block A when it reaches the end of the table.
    (b) The mass of block A is now increased. Explain in words but with specific reference to your calculation in (c) how, if at all, block A's speed at the edge of the table will change.

Lab-Based Free-Response Question with Graphical Analysis

17. In the laboratory, a constant-frequency generator creates standing waves on a rope, as shown in the preceding diagram. The speed of waves on the string is varied by changing the tension in the rope.

    (a) Next to the picture above, draw and label a line indicating one wavelength's distance on the standing wave.
    (b) A student makes the measurements recorded in the table that follows. On the axes provided, label an appropriate scale and graph the wave speed versus the wavelength.

WAVE SPEED (CM/S)	WAVELENGTH (CM)
2,600	45
3,600	59
4,800	76
2,200	38
1,700	32
3,800	62

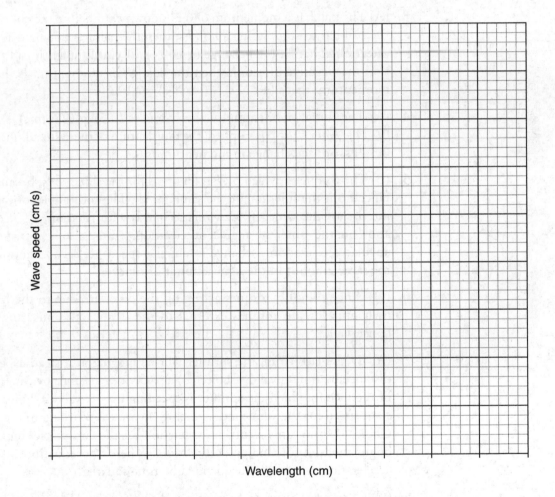

Wave speed (cm/s)

Wavelength (cm)

(c) Use a best-fit line to determine the frequency of the generator.

(d) A different lab group used a frequency generator that provided a frequency of 150 Hz. On the graph, draw and label what a best-fit line for this other group's data might look like.

Solutions for the AP Physics 1 Question Types Assessment

1. (A) The force F_1 is bigger than the force F_2, meaning the net force is in the direction of F_1. Newton's second law says that the direction of acceleration must be the same as the direction of net force.

2. (C) Both F_1 and F_2 apply a counterclockwise torque. So these torques add together to give a net counterclockwise torque. That means the angular speed must change, but not necessarily increase. If the objects were rotating clockwise to start with, then a counterclockwise net torque would slow the angular speed.

3. (A) The ball has a momentum of 6 N·s downward before the collision. In the collision, the ball has to stop, losing all 6 N·s; then the ball has to gain 4 N·s to go in the other direction. That's a total change of 10 N·s. If you'd prefer to call the downward direction the negative direction, then the change in momentum is the final minus initial momentum: that's (+4 N·s) − (−6 N·s) = +10 N·s.

4. (A) The initial momentum of the system is (0.5 kg)(0.6 m/s) + (0.5 kg)(0.8 m/s) = 0.7 N·s. After Cart B is stopped, the momentum of the system is just (0.5 kg)(0.6 m/s) = 0.3 N·s. So, the change in momentum is 0.4 N·s.

5. (B) The torque provided by F is identical for the three rods, because the force and distance from the fulcrum are the same for all. The rods are identical, so their rotational inertias are the same. By Newton's second law for rotation, $\tau_{net} = I\alpha$, because both τ_{net} and I are the same, the angular acceleration α must also be the same for all rods. Then angular acceleration is change in angular speed per second and all change speeds by the same amount in the same amount of time.

6. (D) Since the only torques acting on the diver are due to the forces applied by his own muscles, no torques external to the diver act. Thus, angular momentum is conserved.

7. (A) By the last 10 m, the sprinter will be moving at a steady speed, so just dividing distance by time is valid. Choice (B) also divides distance by time, but during the part of the race in which the sprinter is speeding up—invalid. Choice (C) assumes a constant speed for the whole race, which is incorrect according to the problem statement. Choice (D) not only assumes acceleration for the entire race, but uses a horrendously bogus method for finding acceleration. (There's no equation in the world that says acceleration equals distance divided by time squared.)

8. (C) Choice (A) is wrong because it describes a calculational prediction, not an experimental measurement. Choice (D) determines linear velocity, not angular velocity. Choices (B) and (C) both discuss angular velocity experimentally, but Choice (B) gives the *average* angular velocity, while Choice (C) explicitly describes an instantaneous angular velocity because it is finding the angular displacement over a very short time period when the block is essentially at the position d.

9. (B) Acceleration is the slope of a velocity-time graph. Choices (A) and (D) give speed rather than acceleration. Choice (C) is a distance, not an acceleration.

10. (A) The relevant equation relates change in angular speed per second—that is, angular acceleration—to net torque. That's $\tau_{net} = I\alpha$. The rotational inertia is I, and plotting τ_{net} versus α will give the slope equal to I. Which choice is that? Since the angular speed change was measured over 1 s, $\Delta\omega$ *is* the angular acceleration in this case.

11. (A) The initial kinetic energy is possessed only by the 1-kg cart and is equal to $\frac{1}{2}mv^2 =$ 2 J. We don't know the speed of the carts after the collision, so we have to calculate that via momentum conservation. (Kinetic energy is *not* conserved because the carts stick together—this is an inelastic collision.) The total momentum before collision is the sum of mv for each cart = 2 N·s + 0 = 2 N·s. The total momentum after collision must also be 2 N·s by momentum conservation. (Momentum is *always* conserved in a collision.) The carts' combined mass is 1.5 kg, so the speed must be 1.3 m/s in order to multiply to the 2 N·s total. Kinetic energy of the combined masses is now $\frac{1}{2}(1.5 \text{ kg})(1.3 \text{ m/s})^2 = 1.3$ J.

12. (C) Linear momentum is conserved in all collisions, regardless of the elasticity of the collision. The value of the total momentum will be 2 N·s either way. Kinetic energy was lost in the first collision, because kinetic energy is conserved only in an elastic collision. In the second collision, though, no kinetic energy was lost. Therefore, kinetic energy is bigger after the second collision, but smaller after the first collision.

13. (C) and (D) Choice A is wrong: While it's true that G is much smaller than k, so what? It's the gravitational force of the mass of the stars in the galaxy's center that pulls on the mass of the other stars. Without gravity, stars and planets would not orbit. Choice B is not correct, either. The electric force is, in fact, negligible, but the reasoning presented is wrong. Electric forces dang well can act beyond atomic distances; otherwise, clothes out of the dryer could never experience static cling. The electric force is negligible because electrical interactions can be both attractive *and* repulsive; in sum, virtually all of the attractive and repulsive forces between fundamental particles will cancel. The other statements are correct.

14. (B) and (D) The tire is rotating in the positive direction. Angular velocity and angular momentum are always in the direction of rotation. Since the tire is slowing its rotation, though, angular acceleration must be *opposite* the direction of angular velocity. And by Newton's second law for rotation, net torque is in the same direction as angular acceleration.

15. (A) and (C) Any force that does positive work on the electron will increase the electron's kinetic energy. Positive work means that a component of the force is in the same direction as the electron's motion. Choice (D) is perpendicular to the electron's motion, so that force does no work; choices (A) and (C) all have a component up and to the right, so those forces do work to increase the electron's kinetic energy. Choice B has a component in the direction opposite the electron's motion, so choice B does work to *decrease* the electron's kinetic energy.

16. (a) Treating the two blocks as a single system, the net force is the 10-N hanging weight. The mass of this two-block system is 3 kg, so by $F_{net} = ma$, the system's acceleration is 10 N/3 kg = 3.3 m/s per second.

Now, Block A starts from rest, travels 0.50 m, and accelerates at 3.3 m/s per second. Using the kinematics equation $v_f^2 = v_o^2 + 2a\Delta x$, we get $v_f = \sqrt{(0) + 2(3.3 \text{ m/s/s})(0.50 \text{ m})} = 1.8$ m/s.

(b) The net force on the system does not change, because the 10-N weight is still the same. However, the mass of the system increases. The solution in (a) for the acceleration of the system includes the system's mass in the denominator, so the acceleration of the system will decrease.

Then, since Block A still travels the same distance to the edge of the table from rest, the same kinematics equation as in (a) works, but this time with a smaller acceleration. With that acceleration in the numerator of the final speed equation, the final speed will also decrease.

17. (a)

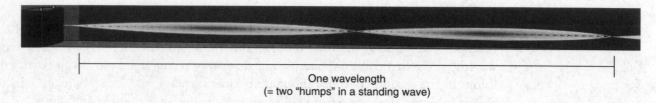

One wavelength
(= two "humps" in a standing wave)

(b)

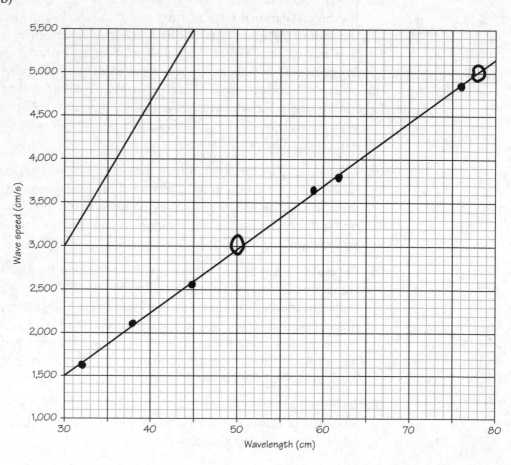

(c) The slope of the line is the frequency of the generator, because the relevant equation is $v = \lambda f$. That matches up with the equation for a line $y = mx$ with the wave speed v on the y-axis and the wavelength λ on the x-axis. The slope value m corresponds with the frequency f.

To calculate the slope, use two points on the line that are *not* data points. I've used the ones that are marked with open circles. Then the slope is the rise over the run (5,000–3,000) cm/s/(78–50) cm. That gives a slope of 71 Hz.

(d) A bigger frequency means a bigger slope. The graph will still go through the origin (because there's no y-intercept term in $v = \lambda f$), and the origin is not pictured in the graph. I've drawn the line such that the slope is about 150 Hz, up 1,500 cm/s on the vertical axis and over 10 cm on the horizontal axis.

STEP 3

Develop Strategies for Success

CHAPTER 6   Strategies to Get the Most Out of Your AP Physics Course

CHAPTER 7   Strategies to Approach the Questions on the Exam

CHAPTER 8   Strategies to Approach the Questions: Free-Response Section

CHAPTER 9   Strategies to Approach the Questions: Multiple-Choice Section

Strategies to Get the Most Out of Your AP Physics Course

IN THIS CHAPTER

Summary: The best way to prepare for the AP Physics 1, Algebra-Based Exam is to make sure you really understand the physics presented in your AP course. This chapter provides strategies you can use to get the most out of your AP Physics 1 class and improve the likelihood that you'll score a 4 or 5 on the test.

Key Ideas

✪ Focus on increasing your knowledge of physics, not on getting a good grade.
✪ Don't spend more than 10 minutes at one time without getting somewhere.
✪ Work with other students.
✪ Ask questions when you don't understand something.
✪ Keep an even temper, and don't cram.

Seven Simple Strategies to Get the Most Out of Your AP Physics Course

Almost everyone who takes the AP Exam has just completed an AP Physics course. *Recognize that your physics course is the place to start your exam preparation!* Whether or not you are satisfied with the quality of your course or your teacher, the best way to start preparing for the exam is by doing careful, attentive work in class all year long.

Okay, for many readers, we're preaching to the choir. You don't want to hear about your physics class. In fact, maybe you're reading this chapter only a few weeks before the exam, and it's too late to do much about your physics class. If that's the case, go ahead to the next chapter, and get started on strategies for the test, not the class.

But if you are reading this a couple of months or more before the exam, we think that you can get even more out of your physics class than you think you can. Read these pieces of time-tested advice, follow them, and we promise you'll feel more comfortable about your class *and* about the AP exam.

1. Ignore Your Grade

This must be the most ridiculous statement you've ever read, right? But it may also be the most important of these suggestions. Never ask yourself or your teacher, "Can I have more points on this assignment?" or "Is this going to be on the test?" You'll worry so much about giving the teacher what he or she wants that you won't learn physics in the way that's best for you. Whether your grade on a class assignment is perfect or near zero, ask, "Did I really understand all aspects of these problems?"

Remember, the AP Exam tests your physics knowledge. If you understand physics thoroughly, you will have no trouble at all on the AP test. But, while you may be able to argue yourself a better grade in your physics class even if your comprehension is poor, the AP readers are not so easily moved.

If you take this advice—if you really, truly ignore your grade and focus on physics—your grade will come out in the wash. You'll find that you got a very good grade after all, because you understood the subject so well. But you *won't care,* because you're not worried about your grade!

2. Don't Bang Your Head Against a Brick Wall

Our meaning here is figurative (although there are literal benefits as well). Never spend more than 10 minutes or so staring at a problem without getting somewhere. If you honestly have no idea what to do at some stage of a problem, *stop*. Put the problem away. Physics has a way of becoming clearer after you take a break.

On the same note, if you're stuck on some piddly algebra, don't spend forever trying to find what you know is a trivial mistake, say a missing negative sign or some such thing. Put the problem away, come back in an hour, and start from scratch. This will save you time in the long run.

And finally, if you've put forth a real effort, you've come back to the problem many times and you still can't get it: relax. Ask the teacher for the solution, and allow yourself to be enlightened. You will not get a perfect score on every problem. But you don't care about your grade, remember?

3. Work with Other People

When you put a difficult problem aside for a while, it always helps to discuss the problem with others. Form study groups. Have a buddy in class with whom you are consistently comparing solutions.

Although you may be able to do all your work in every other class without help, I have never met a student who is capable of solving every physics problem on his or her own. It is not shameful to ask for help. It is not dishonest to seek assistance—as long as you're not copying or allowing a friend to carry you through the course. Group study is permitted and encouraged in virtually every physics class around the globe.

4. Ask Questions When Appropriate

We know your physics teacher may *seem* mean or unapproachable, but in reality, physics teachers do want to help you understand their subject. If you don't understand something, don't be afraid to ask. Chances are that the rest of the class has the same question. If your question is too basic or requires too much class time to answer, the teacher will tell you so.

Sometimes the teacher will not answer you directly but will give you a hint, something to think about so that you might guide yourself to your own answer. Don't interpret this as a refusal to answer your question. You must learn to think for yourself, and your teacher is helping you develop the analytical skills you need for success in physics.

5. Keep an Even Temper

A football team should not give up because they allow an early field goal. Similarly, you should not get upset at poor performance on a test or problem set. No one expects you to be perfect. Learn from your mistakes, and move on—it's too long a school year to let a single physics assignment affect your emotional state.

On the same note, however, a football team should not celebrate victory because it scores a first-quarter touchdown. You might have done well on a test, but there's the rest of the nine-month course to go. Congratulate yourself, and then concentrate on the next assignment.

6. Don't Cram

Yes, we know that you got an "A" on your history final because, after you slept through class all semester, you studied for 15 straight hours the day before the test and learned everything. And, yes, we know you are willing to do the same thing this year for physics. We warn you, both from our and from others' experience: *it won't work.* Physics is not about memorization and regurgitation. Sure, there are some equations you need to memorize, but problem-solving skills cannot be learned overnight.

Furthermore, physics is cumulative. The topics you discuss in December rely on the principles you learned in September. If you don't understand the basic relationships between motion and acceleration, how are you supposed to understand the connection between acceleration and net force, or angular acceleration and net torque?

The answer is to keep up with the course. Spend some time on physics every night, even if that time is only a couple of minutes, even if you have no assignment due the next day. Spread your "cram time" over the entire semester.

> **Exam Tip from an AP Physics Veteran**
> We had a rule in our class: no studying the night before the exam. There was no way to learn something new in the few remaining hours. The goal was to be relaxed and confident about what we did know. In fact, the class all got together for a pool party rather than a study session. And every one of us passed, with three-fourths of us getting 5s.

7. Never Forget, Physics is "Phun"

The purpose of all of these problems, experiments, and exams, is to gain knowledge about physics—a deeper understanding of how the natural world works. Don't be so caught up in the grind of your coursework that you fail to say "Wow!" occasionally. Some of the things you're learning are truly amazing. Physics gives insight into some of humankind's most critical discoveries, our most powerful inventions, and our most fundamental technologies. Enjoy yourself. You have an opportunity to emerge from your physics course with wonderful and useful knowledge, and unparalleled intellectual insight. Do it.

CHAPTER 7

Strategies to Approach the Questions on the Exam

IN THIS CHAPTER

Summary: This chapter contains tips and strategies that apply to both the free-response and the multiple-choice sections of the AP Physics 1, Algebra-Based Exam. First you'll find information about the tools (calculator, equation sheet, and table of information) you can use on the exam and strategies to make the best use of these tools. Then you'll find strategies for dealing with two common types of AP Physics 1 questions: ranking questions and questions about graphs. Both of these types of questions can be found in the multiple-choice section and in the free-response section of the exam.

Key Ideas

✪ Although you can use a calculator on the exam, you should use it only when it's actually required to do a calculation. It won't help you answer the vast majority of the questions on the AP Physics 1, Algebra-Based Exam.

✪ Questions involving a ranking task often require analysis more than calculation. Ranking questions can be multiple choice or free response. For free-response questions, be sure to indicate clearly the order of your ranking, and if two of the items to be ranked are equal, be sure to indicate that, too.

✪ Graph questions are straightforward, because there are only three things you can do with a graph—take the slope, compute the area under the graph, and read one of the axes directly. Pick the right one, and you're golden.

Tools You Can Use and Strategies for Using Them

Like all AP exams, the AP Physics 1, Algebra-Based Exam consists of both multiple-choice questions (Section I) and free-response items (Section II). The three tools discussed below can be used on both sections of the exam.[1] Keep in mind that just because you *can* use these tools, it doesn't necessarily mean you *should*.

Calculator

The rules of acceptable calculators are the same as those on the SAT or the AP math exams—pretty much any available calculator is okay, including scientific and graphing calculators. Just don't use one with a QWERTY keyboard, or one that prints the answers onto paper.[2] You're not allowed to share a calculator with anyone during the exam.

Wait! Don't Touch That Calculator!

Will a calculator actually help you? Not that much. Exam authors are required to write to the specific learning objectives in the "curriculum framework."[3] Of the 140 learning objectives in AP Physics 1 curriculum, *only 21 allow for calculation*! For 119 of 140 learning objectives, something else entirely—qualitative prediction, semiquantitative reasoning, analysis and evaluation of evidence, description of experiment, explanation, etc.—is required. You will need to calculate, but not often.

And even then, the "calculation" on the AP exam usually does not require a calculator. For example, you'll answer in symbols rather than numbers; you'll be asked to do an "order of magnitude estimate" in which only the power of 10 matters[4]; the numbers involved will be simple, like 4×2; or the choices will be so far separated that only one answer will make sense, whether or not you actually carry out the calculation.

> But seemingly every problem we did for class required solving an equation and plugging values into that equation. My teacher assigned us a bunch of problems on the computer, with programs like WebAssign. I could never have done those problems without a calculator!

That's very likely true. Being able to perform calculations is, in fact, a first step toward more difficult physics reasoning. But it's very simple—the AP exam will only sometimes require a calculation. And when it does, a calculator will only sometimes be necessary. So wean yourself off of the calculator. Practice approaching every problem with diagrams, facts, equations, symbols, and graphs. Only use the calculator as a last step in a homework problem. Then you'll be well prepared for the kinds of things you'll see on the AP exam.

The Table of Information

There's no need to memorize the value of constants of nature, such as the mass of an electron or the universal gravitation constant. These values will be available to you on the table of information you'll be given.

---

[1] This represents a change from previous AP Physics exams; you used to have access to a calculator only on the free response.

[2] Does anyone actually use printing calculators anymore?

[3] You can refer to the curriculum framework via the AP Physics 1 portion of the College Board's website, https://www.collegeboard.org.

[4] See Chapter 15 (Gravitation) for specific examples of order-of-magnitude estimates.

The Equation Sheet

A one-page list of many relevant equations will be available to you on both sections of the exam.[5] You will be able to see the official equation sheet ahead of time at the AP Physics 1 portion of the College Board's website (https://www.collegeboard.org).

Wait! Don't Touch That Equation Sheet!

Will the equation sheet actually help you? It won't help you that much. Too often, students interpret the equation sheet as an invitation to stop thinking—"Hey, they tell me everything I need to know, so I can just plug-and-chug through the exam." Nothing could be further from the truth.

First, the equation sheet will likely present most equations in a different form than you're used to, or use different notation than your textbook or your class. So what—you've already memorized the equations on the sheet. It might be reassuring to look up an equation during the exam, just to make sure you've remembered it correctly, which is really the point of the equation sheet. But beware. Use your memory as the first source of equations.

If you must use the equation sheet, *don't go fishing*! If a question asks about a voltage, don't just rush to the equation sheet and search for every equation with a V in it. You'll end up using $\rho = \frac{M}{V}$, where the V means volume, not voltage. You'd be surprised how often misguided students do this. Don't be that person.

> So you're saying I'll be given a calculator and an equation sheet, but neither will be much use. Why would I be given useless items?

Suffice it to say that many years ago, calculation was indeed the most important aspect of an AP Physics exam, so the calculator was indispensable. Back then, students were expected to memorize equations. As calculators became more sophisticated, students began to game the test by programming equations into their calculators, effectively gaining an unfair advantage. So the equation sheet was provided to everyone, negating that advantage. Now that calculation is a far less significant part of AP Physics, the calculator will only rarely be useful. But since you might need it a few times in an exam, it's still allowed.

Strategies for Questions That Involve a Ranking Task

You already know there won't be a lot of straight "calculate this" type of questions. So what kinds of questions will there be? One very different sort of question from the standard textbook end-of-chapter homework problem is the ranking question. It can be found among the multiple-choice questions of Section I or in Section II as a free-response question. Here's an example:

> Cart A takes 5 s to come to rest over a distance of 20 m. Cart B speeds up from rest, covering 10 m in 10 s. And Cart C moves at a steady speed, taking 1 s to cover 50 m. All carts have uniform acceleration. Rank the carts by the magnitude of their acceleration. If more than one cart has the same magnitude of acceleration, indicate so in your ranking.

---

[5]Once again, this is a change for AP Physics 1; the equation sheet used to be for free response only.

Notice that you are emphatically *not* asked to calculate the acceleration of each cart. Usually, a ranking task can be solved more simply with conceptual or semiquantitative reasoning than with direct calculation.

In this example, the conceptual approach is probably best. Acceleration is defined as how quickly an object changes its speed. "Magnitude" of acceleration means ignore the direction of acceleration.[6] Cart C doesn't change its speed at all, so it has the smallest acceleration. Cart A goes farther than Cart B and takes less time to do so. Since both change their speeds from or to rest, Cart A must change its speed more quickly than Cart B.

The multiple-choice ranking tasks will have answer choices formatted as inequalities: A > B > C. If two were equal, then you'd see something like A = B > C, which would mean A and B are equal, but are both greater than C.

In the free-response section, you can format your answer to a ranking task any way that is clear. For example, you could list: "(Greatest) A, B, C (least)." Don't forget to make some notation if two of the choices were equal; circle those two and write "these are equal" or something that is crystal clear.

> **Exam Tip from an AP Physics Veteran**
>
> For some people, semiquantitative and qualitative reasoning is much more difficult than just making calculations. You have every right to start a ranking task with several calculations! Then just rank your answers numerically. Sure, some questions will require you to explain your ranking without reference to numbers, but still, feel free to answer with numbers first, and *then* refer to the equations.

In the example above, you could make the calculation for each cart. Use the kinematics equations detailed in Chapter 10. You can calculate that Cart A has an acceleration of 1.6 m/s per second. Cart B has an acceleration of 0.2 m/s per second. And Cart C has an acceleration of zero.

How could you follow your calculation with a nonnumerical explanation, then? Look at how the equations simplified. For carts A and B, the initial or final speed of zero meant that when you solved in variables for *a*, you got $a = \dfrac{2\Delta x}{t^2}$. Cart A has *both* a bigger distance to travel *and* a smaller time of travel. Cart A's numerator is bigger, denominator smaller, and acceleration bigger than Cart B's.

Strategies for Questions That Involve Graphs

Analyzing data in graphical form will be a skill tested regularly on the AP exam. Good graph questions are straightforward, because there are only *three things you can do with a graph*. Pick the right one, and you're golden.

When you see a graph, the first step must be to *recognize the relevant equation*. In this situation, which equation from the equation sheet relates the *y*- and *x*-variables? I truly mean the equation, not just the units of the axes. Then, the equation will lead you to one of the following three approaches.

---

[6]This can't be determined here anyway, because although Cart A has acceleration opposite its motion, and Cart B has acceleration in the same direction as its motion, we don't know which ways these two carts are moving. But who cares, for this particular question.

What are the three things we can do with this graph?

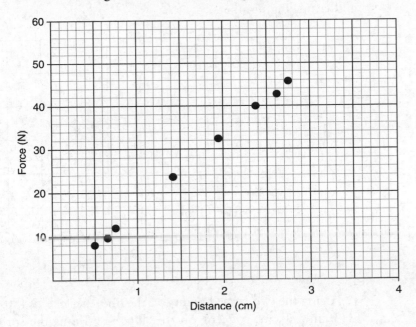

A box sits on a smooth, level surface. The box is attached to a spring. A person pushes the box across the surface, compressing the spring. For each distance the spring compresses, the force applied by the person on the box is measured.

1. Take the Slope of the Graph

You certainly understand that the slope represents the change in the *y*-variable divided by the change in the *x*-variable. But to really understand what the slope of a graph means, you can't just say "it's the change in force divided by the change in distance." Slopes of graphs generally have a physical meaning that you must be able to recognize.

Always start with the relevant equation. If you suspect you're looking for a slope, solve the relevant equation for the *y*-axis variable, then compare the equation to the standard equation for a line: $y = mx + b$. Here, we're talking about the force of a spring and the distance the spring is stretched—that's covered by the equation $F = kx$. *F* and *x* are the vertical and horizontal axes, respectively.

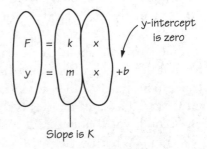

This process of circling the *y*-variable, circling the *x*-variable, and then circling the slope will work with any equation.[7] Here, the slope represents *k*, the spring constant for the spring. So what is the numerical value of the spring constant? You can calculate the

---

[7]And if there's a leftover term with a plus or minus sign, you'll recognize that as the *b*-value, the *y*-intercept of the graph.

slope by drawing a best-fit line, choosing two points on the line that are not data points, and crunching numbers.

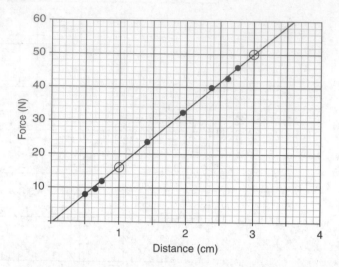

Using the two circled points on the line, the "rise" is (50 N – 16 N) = 34 N. The "run" is (3 cm – 1 cm) = 2 cm. So the slope—and thus the spring constant of the spring—is $\frac{34\,\text{N}}{2\,\text{cm}} = 17\,\text{N/cm}$.

2. Calculate the Area Under the Graph

Always start with the relevant equation. The meaning of the area under the graph is generally found by looking at an equation that *multiplies* the vertical and horizontal axes. Here, that would be force times distance. Sure enough, the equation for work is at work here: $W = F \cdot \Delta x_{\parallel}$. The force applied by the man is parallel to the box's displacement, so the work done by the man is the multiplication of the vertical and horizontal axes. That means that to find the work done, take the area under the graph.

How much work did the man do in compressing the spring 3 cm? Usually, when you're taking an area under a graph on the AP exam, just estimate by breaking the graph into rectangles and triangles. In this case the graph is an obvious triangle. The area of a triangle is (½)(base)(height) = (½)(0.03 m)(50 N) = **0.75 J**. If instead the question had asked for the work done in compressing the spring 2 cm, I'd do the same calculation with 0.02 m and 34 N.

Two things to note about that calculation: First, an area under a graph, like a slope, has units. But the units are *not* "square units" or "square meters." The "area" under a graph isn't a true physical area; rather, it represents whatever physical quantity is found by multiplying the axes. This area represents the work done by the man, so it should have units of joules. Second, you'll note that I used 0.03 m rather than 3 cm in calculating the area under this graph. Why? Because without that conversion, the units of the area would have been newton centimeters (N·cm). I wanted to get the work done in the standard units of joules, equivalent to newton meters (N·m). So, I had to convert from centimeters to meters.

3. Read One of the Axes Directly

Often a question will ask for interpolation or extrapolation from a graph. For example, even though the man never used 60 N of force, how far would the spring compress if he *had* used 60 N? Just extend the best-fit line, as I already did, and read the horizontal axis: 3.6 cm.

Exam Tip from an AP Physics Veteran
When you draw a best-fit line, just lay your ruler down and truly draw the fit as best as you can. Never connect data point-to-point; never force the best-fit line through (0,0); never just connect the first and last data points. In fact, it's best if you extend the line of best fit as far as you can on the graph, so that you can answer extrapolation questions quickly and easily.

Finally, note that many graphs on the AP Physics 1 Exam will include real data, not just idealized lines. Be prepared to sketch lines and curves that seem to fit the general trend of the data.

CHAPTER 8

Strategies to Approach the Questions: Free-Response Section

IN THIS CHAPTER

Summary: The AP Physics 1, Algebra-Based Exam contains question types you probably have not encountered before. This chapter describes types of questions that appear only in the free-response section of the test and the most effective strategies to attack them. Included in this chapter are strategies and advice on how to approach the laboratory question, the qualitative-quantitative translation question, and the free-response questions in general.

Key Ideas

✪ The free-response section contains five questions to be answered in 90 minutes. Included will be a 12-point lab question, a 12-point qualitative-quantitative translation (QQT), and three 7-point short-answer questions. One of these 7-point problems will require a response in paragraph form.

✪ There are six simple strategies and tips you can use when answering the lab question to make sure you get all the points you deserve (see list in this chapter).

✪ If the first part of a QQT requires description, skip it and go to the part of the problem that asks for the calculation. Do the calculation first and then go back to the parts that require qualitative reasoning.

✪ Remember, you're only expected to get about 65 percent of the available points to earn a top score. Don't skip any part of a free-response question—go for partial credit instead. See the list of tips for what to do and what *not* to do to get the most partial credit possible.

✪ The free-response section of the exam is read and graded by humans. At the end of this chapter is a list of tips to best communicate to the reader your understanding of the concepts being tested.

Structure of the Free-Response Section

The free-response section contains five questions to be answered in 90 minutes. The five questions will *not* be all similar in length and style. Instead, the structure will be as follows, but not necessarily in this order:

- One 12-point question posed in a laboratory setting
- One 12-point qualitative-quantitative translation
- Three 7-point short answer questions, one of which will require a response in paragraph form.

The rule of thumb is to spend about two minutes per point answering each question. Start the exam by picking the problem that you can answer the most quickly—that'll probably be one of the shorter, multipart questions. You probably can do that in *less* than two minutes per point. That leaves extra time for a problem that might require more of your efforts.

Included in this chapter are separate discussions of the strategies to use in approaching qualitative-quantitative translation questions and laboratory questions. Other than understanding the appropriate pace and the strategies with which to approach these new question styles, no real extra preparation is necessary. The best thing about the free-response section of the AP Exam is that you've been preparing for it all year long.

Really? I don't recall spending much time in class on test preparation.

But think about all the homework problems you've done. Every week, you probably answer a set of questions, each of which takes a few steps to solve. I'll bet your teacher is always reminding you to show your work carefully and to explain your approach in words.[1] That sounds like what's required on the AP free-response section to me.

How to Approach the Laboratory Question

It is all well and good to be able to solve problems and make predictions using the principles and equations you've learned. However, the true test of any physics theory is whether or not it *works*.

The AP Physics 1, Algebra-Based Exam committee is sending a message to students and teachers that laboratory work is an indispensable part of physics. Someone who truly understands the subject must be able to design and analyze experiments. Not just one, but *two* of the seven "science practices" listed in the curriculum guide refer explicitly to experimental physics. One of the five free-response questions is guaranteed to be posed in a laboratory setting, in addition to some multiple-choice questions that have experimental elements. To conclude, you cannot ignore lab-based questions.

Your ability to answer questions on experiments starts with laboratory work in your own class. It doesn't matter what experiments you do, only that you get used to working with equipment. You should know what equipment is commonly available to measure various physical quantities, and how that equipment works. You should be comfortable describing in words and diagrams how you would make measurements to verify any calculation or prediction you could possibly make in answering a problem.

---

[1] If your teacher *doesn't* expect you to show work clearly and explain your answer in words, do it anyway—that's good AP exam preparation, see?

Now, "laboratory work" doesn't necessarily mean a 10-page, publishable report. Describing an experiment should be a three-sentence, not a three-page, process; analyzing data generally means reading a graph, as discussed in the section above. Here is an example of part of a free-response question that asks for a description of an experimental process.

Sample Laboratory Question

In the laboratory, you are given a metal block, about the size of a brick. You are also given a 2.0-m-long wooden plank with a pulley attached to one end. Your goal is to determine experimentally the coefficient of kinetic friction, μ_k, between the metal block and the wooden plank.

(a) From the list below, select the additional equipment you will need to do your experiment by checking the line to the left of each item. Indicate if you intend to use more than one of an item.

_____ 200-g mass   _____ 10-g mass   _____ spring scale
_____ motion detector   _____ balance   _____ meterstick
_____ a toy bulldozer that moves at constant speed
_____ string

(b) Draw a labeled diagram showing how the plank, the metal block, and the additional equipment you selected will be used to measure μ_k.

(c) Briefly outline the procedure you will use, being explicit about what measurements you need to make and how these measurements will be used to determine μ_k.

Six Simple Strategies and Tips for Answering Descriptive Laboratory Questions

Here are the most effective strategies to use to approach a free-response question that asks for a description of an experimental process like the sample question above.

1. **Follow the directions.** Sounds simple, doesn't it? When the test says, "Draw a diagram," it means you need to draw a diagram. And when it says, "Label your diagram," it means you need to label your diagram. You will likely earn some points just for these simple steps.

Exam Tip from an AP Physics Veteran
On the 1999 AP test, I forgot to label point B on a diagram, even though I obviously knew where point B was. This little mistake cost me several points.

---

2. **Use as few words as possible.** Answer the question, and then stop. You can lose credit for an incorrect statement, even if everything else in your answer is perfect. The best idea is to keep it simple.

3. **There is no single correct answer.** Most of the lab questions are open-ended. There might be four or more different correct approaches. Don't try to "give them the answer they're looking for." Just do something that seems to make sense—you're likely to be right.

4. **Don't assume you have to use all the stuff they give you.** It might sound fun to use a light sensor when determining the net torque on a meterstick, but really? A light sensor?

5. **Don't overthink the question**. You're not supposed to win a Nobel Prize for your work. Free-response problems should never take more than 25 minutes to complete, and usually take much less time. Don't expect to design a subatomic particle accelerator, expect to design a quick measurement that can be done in your classroom.
6. **Write for an audience at the same level of physics as you.** That means, don't state the obvious. You may assume that basic lab protocols will be followed. There's no need to tell the reader that you recorded your data carefully, and you do not need to remind the reader to wear safety goggles.

Answering Lab Questions

Now it's time to pull it all together. Here are two possible answers to the preceding sample question. Look how explicit I am about what quantities are measured, how each quantity is measured, and how μ_k is determined. But, like many physicists, I would have flunked out of art school. Your diagrams don't have to look beautiful because AP readers believe in substance over style. All that matters is that all the necessary components are there in the right places.

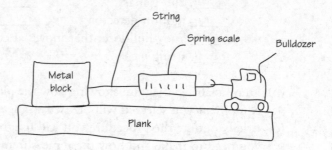

Answer #1

In the laboratory, you are given a metal block, about the size of a brick. You are also given a 2.0-m-long wooden plank with a pulley attached to one end. Your goal is to determine experimentally the coefficient of kinetic friction, μ_k, between the metal block and the wooden plank.

(a) From the list below, select the additional equipment you will need to do your experiment by checking the line to the left of each item. Indicate if you intend to use more than one of an item.

_____ 200-g mass	_____ 10-g mass	✔ spring scale
_____ motion detector	✔ balance	_____ meterstick
✔ a toy bulldozer that moves at constant speed		
✔ string		

(b) Draw a labeled diagram showing how the plank, the metal block, and the additional equipment you selected will be used to measure μ_k.
(c) Briefly outline the procedure you will use, being explicit about what measurements you need to make and how these measurements will be used to determine μ_k.

Use the balance to determine the mass, m, of the metal block. The weight of the block is mg. Attach the spring scale to the bulldozer; attach the other end of the spring scale to the metal block with string. Allow the bulldozer to pull the block at constant speed.

The block is in equilibrium. So, the reading of the spring scale while the block is moving is the friction force on the block; the normal force on the block is equal to its weight. The coefficient of kinetic friction is equal to the spring scale reading divided by the block's weight.

Answer #2

In the laboratory, you are given a metal block, about the size of a brick. You are also given a 2.0-m-long wooden plank with a pulley attached to one end. Your goal is to determine experimentally the coefficient of kinetic friction, μ_k, between the metal block and the wooden plank.

(a) From the list below, select the additional equipment you will need to do your experiment by checking the line to the left of each item. Indicate if you intend to use more than one of an item.

   ✔ 200-g mass   **✔** 10-g mass   _____ spring scale
(several)   (several)

   ✔ motion detector   **✔** balance   _____ meterstick

   _____ a toy bulldozer that moves at constant speed

   ✔ string

(b) Draw a labeled diagram showing how the plank, the metal block, and the additional equipment you selected will be used to measure μ_k.

(c) Briefly outline the procedure you will use, being explicit about what measurements you need to make and how these measurements will be used to determine μ_k.

Determine the mass, m, of the block with the balance. The weight of the block is mg. Attach a string to the block and pass the string over the pulley. Hang masses from the other end of the string, changing the amount of mass until the block can move across the plank at constant speed. Use the motion detector to verify that the speed of the block is as close to constant as possible.

The block is in equilibrium. So, the weight of the hanging masses is equal to the friction force on the block; the normal force on the block is equal to its weight. The coefficient of kinetic friction is thus equal to the weight of the hanging masses divided by the block's weight.

The Qualitative-Quantitative Translation (QQT)

While physics is *not* about numbers (see Chapter 2), physicists routinely plug values into relevant equations to produce numerical predictions such as "the reading on the scale will be 550 N." In AP language, that's *quantitative* reasoning.

Just as routinely, though, physicists use equations as the basis for a less-specific conceptual prediction. Without using any numbers, it's usually possible to look at the features of an equation—what variables are in the numerator or denominator, what's squared or square rooted, what is added or subtracted—and determine something like "the reading on the scale will increase." In AP language, that's *qualitative* reasoning.

One of the five free-response items on the AP Physics 1 Exam is called a "qualitative-quantitative translation question," abbreviated as QQT. It will generally ask you first for qualitative reasoning, next for quantitative reasoning, and *then* it will expect you to explain in words how the different aspects of your solutions relate to each other. If you think of algebra as the language of physics, the QQT will ask you to translate from algebra to English.

An Important Strategy for Solving QQTs

The fact is, by the time they're through a full year of physics, most first-time physics students are far more comfortable with numerical calculation than they are with verbal description of physics concepts. Partly that's because most physics classes emphasize calculation more than writing; partly it's because people are taught to calculate in elementary school but are rarely asked to write in words how and why a calculation worked.

Play to your strengths. If the first part requires description, skip it and go to the part of the problem that asks for the calculation. Do the calculation. Write out every step carefully, annotating your solution with words explaining the point of each step. Then go back to the parts that require qualitative reasoning. Now, since you've approached the situation the "easier" way, you should have a good clue as to why you did the mathematical steps you did.

The actual qualitative-quantitative translation question will be one of the longer questions on the exam, a 12-point question. Usually only part of the question will directly ask about translating calculation into words, and vice versa. Here's an example of the actual translation portion of such a question.

Sample QQT

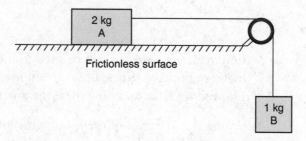

Two blocks are connected over a light, frictionless pulley, as shown. Block A of mass 2 kg is on a frictionless surface; Block B of mass 1 kg hangs freely. The blocks are released from rest. (A) After the blocks are released, is the tension in the rope greater than, less than, or equal to the weight of Block B? Justify your answer in words, without equations or calculations. (B) Calculate the tension in the rope after the blocks are released.

Answering the QQT

Understand that you might see several other parts, say, asking for a free-body diagram of each block, or asking about the behavior of the blocks after Block B hits the ground, or whatever. For now, though, let's just focus on the true translation in Parts (A) and (B).

Even though Part (A) specifically forbids equations and calculations, try approaching it like a calculation problem, anyway. Part (B) asks for the actual value of the tension in the rope. Take care of that first.

Treating the two blocks as a single system, the net force is the 10-N hanging weight. The mass of this two-block system is 3 kg, so by $F_{net} = ma$, the system's acceleration is 10 N/3 kg = 3.3 m/s per second.

Next, consider a system as *just* Block A. The only force acting on Block A is the tension in the rope, so that's the net force. The tension then is the mass of Block A times its acceleration, (2 kg)(3.3 m/s per second) = **6.6 N.**

There's the answer to Part (B), with a nicely explained calculation. What about Part (A)? Well, we at least know the answer from our calculations—the tension is *less than* the 10-N weight of Block B. But now we need to justify the answer without doing the calculation. How do we do that? Describe why the calculation came out as it did, without doing the actual calculation. Be specific about what values are bigger, smaller, or the same throughout the calculation.

Correct Answer #1: "Block A has the same acceleration as the two-block system but less mass than the whole system. Block A experiences a smaller net force than the whole system. The tension is the net force on Block A; the weight of Block B is the net force on the whole system. The tension is less than Block B's weight."

To take a different approach, we know that force problems are best approached with free-body diagrams before plugging in any numbers to equations. An alternate explanation might be to look at a free-body diagram of just Block B.

Correct Answer #2: "The tension pulls up on Block B, and the weight pulls down. Since Block B has a downward acceleration—it speeds up and moves down—the downward forces must be bigger than the upward forces. So the tension is less than the weight."

In either case, you're using words to describe the way you would solve the problem if you were doing a calculation. That's translating from quantitative reasoning to qualitative reasoning—the very definition of a QQT.

What Do the Exam Readers Look For?

The key to doing well on the free-response section is to realize that, first and foremost, these problems test your understanding of physics; next, they test your ability to communicate that understanding. The purpose of the free-response questions is not to see how good your algebra skills are, or how many fancy-sounding technical terms you know, or how many obscure ideas you can regurgitate. You already know from Chapter 1 that the free-response section of the AP Physics 1 Exam is graded by human readers, not a computer. All I'm going to do in this section is give you some important suggestions about how you can best communicate to the reader that you understand the concepts being tested.

All free-response questions are graded by physics teachers who must carefully follow a "rubric" for each question. A rubric is a grading guide—it specifies how points are awarded, including the elements of an answer necessary for both full and partial credit.

You Cannot "Game" the Rubric

It's tempting to try to find that "One Weird Trick" that will guarantee you an extra point or two on the exam. You and your teacher might look at previous years' rubrics[2] and think you see a pattern. But I warn you: Tricks don't work to solve AP Physics 1 problems.

Each rubric is unique. It's created originally by the author of the test question, revised by the exam development committee, adjusted by table leaders based on actual student responses and feedback from colleagues, and finalized mere hours before the reading begins.

---

[2] You can find rubrics to old AP Physics B and C exams on the College Board's AP Central website. Although the principles of grading to a rubric will not change with the switch to AP Physics 1, the style of the rubrics will likely be different in future years.

The College Board does not have hard and fast rules about how a rubric should be written. Each table leader has wide latitude to adjust the rubric so that it awards credit for good physics, and so it does *not* award credit for bad physics.

Rubrics Provide Ample Opportunity for Partial Credit

Recall that you're only expected to get about 65 percent of the available points to earn a top score. You're not supposed to answer every problem perfectly; you're expected to communicate as much physics understanding as possible. Rubrics are designed to award partial credit for answers that are correct but incomplete or that are essentially correct with only minor mistakes.

Thus, your strategy should always be to make a reasonable attempt at each part of every problem. Don't fret that you don't know how to do everything perfectly; just give the best answer you can, and expect to earn credit in proportion to how well you're explaining what you *do* know.

Here are some hints regarding partial credit:

- If you can't solve Part (a) of a multipart problem, don't skip parts (b) through (e)! Sometimes you'll get everything else perfect if you just move along.
- If the answer to Part (b) depends on the answer to Part (a), it's okay to say "I didn't get Part (a), but pretend the answer was 25 m/s." As long as your answer isn't absolutely silly,[3] you will get full or close to full credit. Rubrics are generally designed not to penalize the same wrong answer twice.
- If you get a right answer using a shortcut—say, by doing a calculation in your head, or just remembering a demonstration in your class—you're likely to get credit for your answer but not for the reasoning behind your answer.
- However, if you get a *wrong* answer using a shortcut, you're not likely to get any credit at all. AP readers can read only what's written on your test. They cannot read your mind, and they are not allowed to assume that you know what you're talking about. The moral of the story is this: Communicate with the readers so you are sure to get all the partial credit you deserve.

Exam Tip from an AP Physics Veteran
When I first started physics class, I became frustrated that I didn't get full credit on questions where I thought I understood the right answer. I tried to argue with my teacher after a test: "Here, let me tell you what I was trying to say, so you can give me these points that I deserve." My teacher told me, "So, are you allowed to go to Kansas City where the AP Exam is graded, and go along with your test from reader to reader to tell them what you really meant?" From then on, I started being more careful to explain my answers thoroughly on homework problems and on tests. I got an easy 5 on the AP Exam.

You should also be aware of some things that will *not* get you partial credit:

- You *cannot* earn partial credit if you write multiple answers to a single question. If AP readers see that you've written two different answers, they are instructed to grade the one that is incorrect, even if the other is correct. If you're not sure of the answer, you can't hedge your bets.

---

[3] Like saying that a person not named "Clark Kent" was running 25 m/s.

- You *cannot* earn "extra" credit. Readers are not allowed to say, "Wow, that's the most complete answer I've ever heard. Here's +1." Don't include unnecessary information. It won't help, and if you make a misstatement, you will actually lose points. Answer the question fully, and then stop.

Final Advice About the Free-Response Questions

Here are some final tips and advice regarding the free-response section of the test:

- Annotate any calculations with words. Explain why you're using the equation you're using, and show the values you're plugging in.
- If you don't know exactly how to solve part of a problem, it's okay to explain your thinking process as best you can. For example, "I know the centripetal force points toward the center of the satellite's orbit, and I know it's a gravitational force. But the centripetal acceleration cannot be calculated because I don't know the value of this centripetal force." Such an answer might earn partial credit, even if you were supposed to do a calculation.
- Don't write a book. Even a question that asks for an essay-style response should be answered in a couple of short paragraphs, not a few long pages.
- If you make a mistake, cross it out. If your work is messy, circle your key points or your final answer so that it's easy to find. Basically, make sure the readers know what you want them to grade and what you want them to ignore.
- If you're stuck on a free-response question, try another one. Question 5 might well be easier for you than Question 1. Get the easy points first, and only after that, try to get the harder points with your remaining time.
- Put units on every numerical answer.
- Don't be afraid to draw—diagrams, graphs, or whatever—in response to a question. These may be useful elements of an explanation, especially on the occasions when you're forbidden from using numbers or equations. Be sure to label diagrams and graphs.
- If your approach is so complicated that it's not doable in 15 to 20 minutes with minimal calculator use, you're doing it wrong. Look for a new way to solve the problem, or just skip it and move on.

CHAPTER 9

Strategies to Approach the Questions: Multiple-Choice Section

IN THIS CHAPTER

Summary: Even the multiple-choice section of the AP Physics 1, Algebra-Based Exam contains types of questions you probably have not encountered before. This chapter contains strategies and tips to attack the multiple-choice questions, including the new "multiple-correct" questions.

Key Ideas

✪ The multiple-choice section contains 50 questions, which require the same sorts of in-depth reasoning as the free-response questions.

✪ There's no guessing penalty, so be sure you mark an answer to every item.

✪ Some questions will be identified as "multiple-correct" questions. For these, you need to choose exactly two answers. There's no partial credit—you have to mark *both* of the correct answers, and *neither* of the wrong answers, to earn your point.

✪ To prepare for the test, practice pacing yourself using multiple-choice problems, and for the ones you miss, write explanations or justifications for the correct answers. This will truly allow you to learn from your mistakes.

Multiple-Choice Questions

The multiple-choice section comes first. You have 90 minutes to answer 50 questions. Each question will include four choices (not five, as on the old AP Physics B Exam). There's no guessing penalty, so be sure you mark an answer to every item.

Before even thinking about strategies, understand that multiple-choice questions require the same sort of in-depth reasoning as do the free-response questions. The only difference is that you're not expected to write out justification for your answers. You will see graphs, calculations, lab questions, ranking tasks, and explanations—everything that you're used to from your physics class and everything that you could see on the free-response section. Don't try to breeze through or to "game" the test. Just answer carefully, justifying each answer in your mind with a fact, equation, or calculation.

You have nearly two minutes per multiple-choice question. Some answers will be obvious; knock these out quickly, so that you leave more time to look at the more complicated questions.

Multiple-Correct: A New Question Type

Probably every multiple-choice test you've ever taken has asked you to pick the *best* answer from four or five choices for each problem. Well, that's gonna change.

A subsection of five questions on the multiple-choice section will include "multiple-correct" questions. For these, you will be asked to choose exactly two answers. There's no partial credit here—you have to mark *both* of the correct answers, and *neither* of the wrong answers, to earn your point.

These multiple-correct items replace the old-style questions with roman numerals I, II, and III. Also note that AP Physics 1, Algebra-Based multiple-choice questions will never ask something like, "Which of the following is *not* correct?" The College Board decided that multiple-correct questions were far easier to read and understand than the more confusing styles of past questions. This is supposed to be a physics test, after all, not a set of Hobbit-style riddles.

You will always know which questions are "multiple correct" as opposed to "single answer." The only change to your approach on multiple-correct questions should be that you can't eliminate the three wrong answers to find the one right answer—you have to consider each of the four choices on its own merits.

> **Example:** A cart on a track is moving to the right and has been moving to the right for the last 2 s. Choose the correct statements about physical quantities related to the cart. Choose two answers.
> (A) The cart's displacement vector for the 2 s of motion is directed to the right.
> (B) The cart's instantaneous velocity vector is directed to the right.
> (C) The net force acting instantaneously on the cart is directed to the right.
> (D) The cart's average acceleration over the 2 s of motion is directed to the right.

This question requires just a bit more mental discipline than a standard multiple-choice item. Start by looking at choice (A). Displacement means, where does the cart end up in relation to its starting point? The cart ended up to the right of where it started, so the displacement is to the right. Mark choice (A).

In standard multiple-choice questions, you'd be done, dusted, and reading the next question by now. Not here, though. Keep reading. (B) The direction of instantaneous velocity is simply the direction that the cart is moving in right now. That's also to the right. Also mark choice (B).

Then look at (C) and (D). The direction of acceleration cannot be determined unless we know whether the cart is speeding up or slowing down. Net force is always in the direction of acceleration; because the acceleration direction is unknown, so is the net force direction. Don't mark choices (C) or (D).

Preparing for the Multiple-Choice Section of the Test

Pacing Yourself

Your physics teacher has access to all sorts of multiple-choice questions that are at least somewhat similar to what you'll see on the AP Physics 1 Exam. The 2004 and 2009 AP Physics B exams were released for use in physics classes, an additional practice AP Physics B Exam was released to teachers, several SAT II: Physics tests have been made available for classroom use, and a sample AP Physics 1 test was released in the summer of 2014. Probably, you've seen questions from these sources on your in-class tests, your semester exam, or in practice packets. Use your in-class multiple-choice questions as preparation for the real AP Physics 1 Exam.

Whatever you do, *don't* look at a big set of multiple-choice questions at your leisure, trying them and looking up the answers. Instead, take a set of multiple-choice questions as an authentic test. The real exam gives 50 questions in 90 minutes; so you should attempt 25 questions in 45 minutes, for example, or 16 questions in 30 minutes.

Do enough of these practice tests, and you'll learn the correct pace. Are you getting to all the questions? If not, you're going to need to decide your strengths and weaknesses. You'll figure out before the real exam which types of problems you want to attempt first. You'll learn through practice when you're lingering too long on a single problem.

Test Corrections—The Best Way to Prepare for the Test

When you're done with a practice test, *don't* look up the solutions yet. Have someone else check your answers. For each problem you got wrong, talk to a friend about the problem. Either alone or with your friend's help, write out a justification for the correct answer as if this were the free-response exam. Take about half a page for each justification. Have someone else, like your teacher, check your work again, marking the justifications that still are incomplete or incorrect. Do these *again*, even if you need a lot of help.

The point is, in preparing for the exam it is way too easy to fool yourself into thinking you understand a problem if you look at the solutions too soon. That's why your physics class gives tests—not just to put a grade on the report card, but to give you authentic feedback on what you know, and what you need to work on. If you are thorough and careful in correcting your in-class and practice tests, you will find your multiple choice scores improving rapidly. (But if you show up on exam day without ever taking a practice exam and doing corrections, you will be doing *a lot* of guessing.)

Final Strategies for the Multiple-Choice Section

Here are some final strategies and advice that will help you score higher on the multiple-choice section of the AP Physics 1 Exam:

- The multiple-choice questions will not necessarily start easy and get harder (unlike those on the SAT which start easy and get hard). If you suspect from your practice that you may be pressed for time, know that the problems on your strong topics may be scattered throughout the exam. Problem 50 may be easier for you than problem 5, so pace yourself so that you can get through the whole test and at least get all the easy answers right.

- If you don't see a direct approach to answering a question, look at the choices. They might give you a hint. For example, one of the choices might be way too fast for a car to be moving, or one of the explanations of a concept might contain an obvious error. Then, great, you can eliminate the obvious "stupidicism," guess from the other choices, and move on. Don't dwell on a single question.

- There are no trick questions on this exam. An answer choice cannot "deny the stem." This means, if a problem asks "Which of the following evidence shows that momentum is conserved in this situation?" one answer choice will not be "Momentum wasn't even conserved."

- Never try to "game" the test. Don't approach a question thinking, "What do they[1] want me to say?" There is no trick, no ulterior motive in the question. Just show your physics knowledge, or take your best guess.

- Often a good guess is evidence of good physics instincts. Don't be afraid to make a guess based on your intuition. If you've been practicing physics problems appropriately and you've been correcting your practice tests, then your instincts will be finely honed by exam time. What you call a "guess" may well be much better than a shot in the dark.

---

[1]Who do you mean by "they" anyway? The test is written by some of the country's best physics teachers. Their job is to test whether you know physics, not to show how clever they are, not to embarrass you, and not to play "gotcha!"

STEP 4

Review the Knowledge You Need to Score High

CHAPTER 10   Motion in a Straight Line

CHAPTER 11   Forces and Newton's Laws

CHAPTER 12   Collisions: Impulse and Momentum

CHAPTER 13   Work and Energy

CHAPTER 14   Rotation

CHAPTER 15   Gravitation

CHAPTER 16   Electricity: Coulomb's Law and Circuits

CHAPTER 17   Waves and Simple Harmonic Motion

CHAPTER 18   Extra Drills on Difficult but Frequently Tested Topics

CHAPTER 10

Motion in a Straight Line[1]

IN THIS CHAPTER

Summary: The entire goal of motion analysis is to describe, calculate, and predict where an object is, how fast it's moving, and how much its speed is changing. In this chapter you'll review two separate approaches to make these predictions and descriptions: graphs and algebra.

Definitions

KEY IDEA

✪ The cart's **position** (x) tells where the cart is on the track.
✪ The cart's **speed** (v) tells how fast the cart is moving.[2]
✪ **Acceleration** (a) tells how much the object's speed changes in one second. When an object speeds up, its acceleration is in the direction of its motion; when an object slows down, its acceleration is opposite the direction of its motion. $\rightarrow v=+$ & $a=+$ or $v=-$ & $a=-$
✪ **Displacement** (Δx) tells how far the object ends up away from its starting point, regardless of any motion in between starting and ending positions.
✪ The graphical analysis of motion includes **position-time graphs** and **velocity-time graphs**. On a position-time graph, the slope is the object's speed, and the object's position is read from the vertical axis. For velocity-time graphs, the speed is read from the vertical axis, and the slope is the object's acceleration. $\rightarrow$ derivative $\leftarrow$

---

[1] We're discussing motion with constant acceleration; that covers pretty much all motion in a straight line in AP Physics 1 except things attached to springs.

[2] Strictly speaking, speed is the magnitude—the amount—of the velocity vector. Velocity tells how fast something moves, as well as in which direction it moves. I tend to use "speed" and "velocity" interchangeably, especially in this unit; the distinction between the two is not important here.

✪ The **five principal motion variables** are:
v_0 initial velocity
v_f final velocity
Δx displacement
a acceleration
t time

✪ In any case of accelerated motion, when three of the five principal motion variables are known, the remaining variables can be solved for using the kinematic equations.

✪ **Free fall** means no forces other than the object's weight are acting on the object.

✪ A **projectile** is an object in free fall, but it isn't falling in a straight vertical line. To approach a projectile problem, make *two* motion charts: one for vertical motion and one for horizontal motion.

Introduction to Motion in a Straight Line

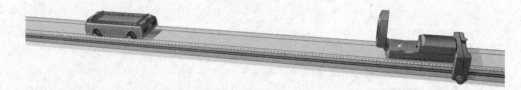

Pretty much all motion problems can be demonstrated with a cart on a track, like in the diagram above. The motion detector can read the location of the cart up to 50 times each second. This detector can make graphs of position or velocity versus time.

The entire goal of motion analysis is to describe, calculate, and predict where the cart is; how fast it's moving; and how much its speed is changing. You'll use two separate approaches to make these predictions and descriptions: graphs and algebra.

Graphical Analysis of Motion

Before you start any analysis, tell yourself which kind of graph you're looking at. The most common mistake in studying motion graphs is to interpret a velocity-time graph as a position-time graph, or vice versa.

Position-Time Graphs

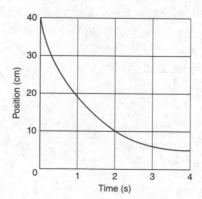

Example 1: The preceding position-time graph represents the cart on the track. The motion detector is located at position $x = 0$; the positive direction is to the left.

An AP Exam question could ask all sorts of questions about this cart. How should they be approached? Use these facts, and reason from them.

FACT: In a position-time graph, the object's position is read from the vertical axis.

Look at the vertical axis in Example 1. At the beginning of the motion, the cart is located 40 cm to the left of the detector.[3] After 2 s, the cart is located 10 cm left of the detector. Therefore, in the first 2 s of its motion, the cart moved 30 cm to the right.

FACT: In a position-time graph, the object's speed is the slope of the graph. The steeper the slope, the faster the object moves. If the slope is a front slash (/), the movement is in the positive direction; if the slope is a backslash (\), the movement is in the negative direction.

Wait a second, how can I tell the slope of a curved graph? Just look at how the graph is sloped at one specific place on the graph. If I want to know how fast the object is moving after 2 seconds of motion, I take the slope drawn in the following figure.

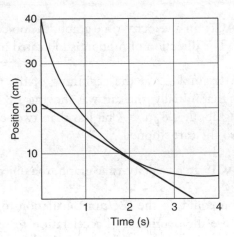

Mathematically, this means you divide (change in y-value)/(change in x-value) to get a speed that's a bit more than 6 cm/s. The exact calculation is not something to obsess over. It's more important to think in terms of comparisons.

You're not often going to be asked "Calculate the speed of the cart at $t = 2$ s." Instead, you'll be asked to describe the motion of the cart in words and to justify your answer. When you describe motion, use normal language that your grandparents would understand. Avoid technical terms like "acceleration" and "negative." Justify your answer with direct reference to the facts.

Referring back to Example 1, because the slope was steeper at earlier times and shallower at later times, the cart must be slowing down. The car is moving to the right the whole time, because the slope is always a backslash.

---

[3]Why not 40 cm to the right of the detector? Because the position value is +40 cm and the positive direction is left.

Velocity-Time Graphs

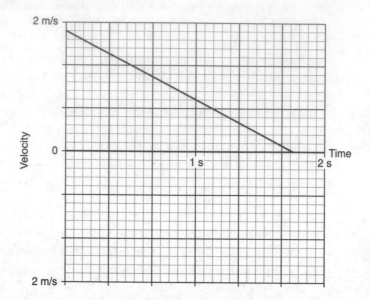

Example 2: The preceding velocity-time graph represents a different cart on the track. The positive direction is to the left.

FACT: In a velocity-time graph, the object's speed is read from the vertical axis. The direction of motion is indicated by the sign on the vertical axis.

In Example 2, at the beginning of the motion, the vertical axis reads 1.8 m/s. This means that initially, the cart was moving 1.8 m/s to the left. After one second, the cart was moving about 0.8 m/s. A bit less than two seconds into the motion, the vertical axis reads zero, so the cart stopped.

FACT: In a velocity-time graph, the object's acceleration is the slope of the graph.

You could do the rise/run calculation to find the amount of the acceleration, or you could use the definition of acceleration to see that the object lost 1 m/s of speed in one second, making the acceleration 1 m/s per second.[4] The cart in Example 2 was slowing down and moving to the left. When an object slows down, its acceleration is opposite the direction of its motion; this cart has an acceleration to the right.

The Mistake

Acceleration is *not* the same thing as speed or velocity. Speed says how fast something moves; acceleration says how quickly speed changes. Acceleration doesn't say anything about which way something is moving, unless you know whether the thing is speeding up or is slowing down.

---

[4]Textbooks and problems on the AP Exam will write this as 1 m/s². Well, that's silly—what the heck is a "second squared," anyway? When you see that notation, read it as "meters per second per second." I suggest you always write the units of acceleration as "m/s per second." Then you'll be far less likely to make "The Mistake."

Someone who says, "This car has an acceleration of 4 m/s per second, so it is moving at about a jogging pace," has made "The Mistake." The car is speeding up or slowing down by 4 m/s every second; the car could well be an Indy 500 racecar traveling 94 m/s right now, but only 90 m/s a second later.

Someone who says "The cart in Figure 2 has a negative acceleration, so it is moving to the right" has made "The Mistake." An acceleration to the right means either speeding up and moving right, or slowing down and moving left. While the cart's acceleration is negative—after all, the slope of the line is a backslash—the car was slowing down, making the velocity's direction opposite the acceleration's direction. The acceleration is right, and the velocity is left.

It takes a lot of practice to avoid "The Mistake." Just continually remind yourself of the meaning of acceleration (how much an object's speed changes in one second), and you'll get there.

FACT: The object's displacement is given by the area between the graph and the horizontal axis. The location of the object can't be determined from a velocity-time graph; only how far the object ended up from its starting point can be determined.

To find how far the cart in Example 2 moved, take the area of the triangle in the graph,[5] giving about 1.6 m. Since the cart's velocity as read from the vertical axis was positive that whole time, and the positive direction is left, the cart ended up 1.6 m left of where it started. But exactly where it started, no one knows.

Algebraic Analysis of Motion

Example 3: A model rocket is launched straight upward with an initial speed of 50 m/s. It speeds up with a constant upward acceleration of 2.0 m/s per second until its engines stop at an altitude of 150 m.

Sometimes you'll be asked to analyze motion from a description and not a graph. Start your analysis by defining a positive direction and clearly stating the start and the end of the motion you're considering. For example, take the upward direction as positive,[6] and consider from the launch to when the engines stop.

Next, *make a chart* giving the values of the five principal motion variables. Include a plus or minus sign on every one (except time—a negative time value means you're in a Star Trek–style movie). If a variable isn't given in the problem, leave that variable blank.

The Five Principal Motion Variables for Your Chart

v_0 initial velocity
v_f final velocity
Δx displacement
a acceleration
t time

---

[5]The area of a triangle is (1/2) base × height.

[6]Could I have called the downward direction positive? Sure. Then signs of displacement, velocity, and acceleration would all be switched.

For Example 3, the chart looks like this:

v_0	+50 m/s
v_f	
Δx	+150 m
a	+2.0 m/s per second
t	

The acceleration is positive because the rocket was speeding up; therefore, acceleration is in the same direction as the motion, which was upward. Upward was defined as the positive direction here.

FACT: In any case of accelerated motion when three of the five principal motion variables are known, the remaining variables can be solved for using the kinematic equations.

In Example 3, we know three of the five motion variables; therefore, we can find the others, and the physics is *done*.

Whoa there. Um, how is the physics "done"? Don't we have to plug the numbers in to the kinematic equations, which incidentally you haven't mentioned yet?

Well, remember the AP Physics 1 revolution: While you will occasionally be asked, say, to calculate how much time the engines run for, you'll just as often be asked something that doesn't involve calculation. For example, "Is it possible to determine the running time of the engines?" Or, "When the engines have run for half of their total run time, is the rocket at a height greater than, less than, or equal to 75 m?"[7]

More to the point, actually doing the math here is, well, a *math* skill not a physics skill. As long as your answers are reasonable—a model rocket will likely burn for a few seconds, not a few thousands of seconds—the exam is likely to award close to full, or sometimes even full, credit for a correct chart and for recognizing the correct equation to use.

FACT: To calculate the missing values in a motion chart, use the three kinematic equations listed as follows. Choose whichever equation works mathematically. Never solve a quadratic equation. If the math becomes overly complicated, try solving for a different missing variable first.

Kinematic Equations

1. $v_f = v_0 + at$

2. $\Delta x = v_0 t + \dfrac{1}{2} at^2$

3. $v_f^2 = v_0^2 + 2a\Delta x$

---

[7]The answer is *less than* 75 m. The rocket is speeding up throughout the time when the engines burn. In the second half of the burn time, the rocket is (on average) moving faster, and so it covers more distance.

Continuing with Example 3, we can use equation (3) to solve for the final velocity of the rocket[8]; this is about 56 m/s.[9] Then we can use equation (1) to get the time before the engines shut off, which is 3 seconds. If you had tried to use equation (2) to solve for time, you would have gotten a quadratic; that's why I said to use equation (3) and then (1).

Objects in Free Fall

FACT: When an object is in free fall, its acceleration is 10 m/s per second[10] toward the ground. "Free fall" means no forces other than the object's weight are acting on the object.

Let's do more with the rocket in Example 3. When the engines stop, the rocket is moving upward at 56 m/s. The rocket doesn't just stop on a dime. It keeps moving upward, but it slows down, losing 10 m/s of speed every second.

Try making a chart for the motion from when the engines stop to when the rocket reaches the peak of its flight. We'll keep the positive direction as upward.

v_0	+56 m/s
v_f	0 (The peak of flight is when the object stops to turn around.)
Δx	
a	−10 m/s per second (There is negative acceleration because free-fall acceleration is always *down*.)
t	

Three of the five variables are known so the physics is *done*.

Now, be careful that you keep grounded in what's physically happening, not in the algebra. For example, you might be asked for the maximum height that the rocket in Example 3 reaches.

Good job recognizing that you need equation (3) to solve for Δx. Dropping the units during the calculation, that gives $0^2 = 56^2 + 2(-10)(\Delta x)$. Solving with a calculator you get about 160 m for Δx.

Wait a second! Think what the 160 m answer means—that's the distance the rocket goes between when the engines stop and when the rocket reaches its highest height. That's not the height above the ground, just the additional height after the engines stop! The actual maximum height is this 160 m, plus the 150 m that the rocket gained with its engines on, for a total of 310 m.

If you were blindly plugging numbers into equations, you would have totally missed the meaning behind these different distances. The AP Physics 1 Exam will repeatedly ask targeted questions that check to see whether you understand physical meaning. Calculation? Pah. In comparison, it's not so important.

Projectile Motion

A *projectile* is defined as an object in free fall. But this object doesn't have to be moving in a straight line. What if the object were launched at an angle? Then you treat the horizontal and vertical components of its motion separately.

---

[8]This calculation requires a calculator, of course: $v_f^2 = (50 \text{ m/s})^2 + 2(2 \text{ m/s per second})(150 \text{ m})$. See, here's another reason you're probably not going to have to actually carry out this math—you're not likely to need a calculator more than a few times on the entire exam.

[9]Not 55.67764363 m/s. Don't make me ask your chemistry teacher to talk to you about significant figures again. Just use two or three figures on all values, and we'll all be happy.

[10]No, stop it with the 9.8 m/s per second. The CollegeBoard is very clear that you can and should use a free-fall acceleration of 10 m/s per second. Really. Limited calculator use, remember?

Example 4: A ball is shot out of a cannon pointed at an angle of 30° above the horizontal. The ball's initial speed is 25 m/s. The ball lands on ground that is level with the cannon.

FACT: A projectile has no horizontal acceleration and so moves at constant speed horizontally. A projectile is in free fall, so its vertical acceleration is 10 m/s per second downward.

To approach a projectile problem, make *two* motion charts: one for vertical motion and one for horizontal motion.

FACT: To find the vertical component of a velocity at an angle, multiply the speed by the sine of the angle. To find the horizontal component of a velocity at an angle, multiply the speed by the cosine of the angle. This always works, as long as the angle is measured from the horizontal.

Here are the two charts for this ball's motion in Example 4. Consider up and right to be the positive directions. Let's consider the motion while the ball is in free fall—that means, starting right after the ball was shot, and ending right before the ball hits the ground. Note that the initial vertical velocity is $(25 \text{ m/s})(\sin 30°) = 13$ m/s. The initial horizontal velocity is $(25 \text{ m/s})(\cos 30°) = 22$ m/s. You needed to use your calculator to get these values.

Vertical	**Horizontal**
v_0 + 22 m/s	v_0 + 13 m/s
v_f	v_f
Δx 0	Δx
a −10 m/s per second	a 0
t	t

Two entries here are tricky. Remember that displacement only means the distance traveled start to end, regardless of what happens in between. Well, this ball landed on "level ground." That means that the ball ends up at the same vertical height from which it was shot; it didn't end up any higher or lower than it started. Thus, vertical displacement is zero.

Second, the final vertical velocity is unknown, not zero. Sure, once the ball hits the ground it stops; but then it's not in free fall anymore. The "final" velocity here is the velocity in the instant before the ball hits the ground.

FACT: The horizontal and vertical motion charts for a projectile must use the same value for time.

The vertical chart is completely solvable, because three of the five variables are identified. Once the time of flight is calculated from the vertical chart, that time can be plugged into the horizontal chart, and *voila*, we have three of five horizontal variables identified; the chart can be completed.

› Practice Problems

Note: Extra drills on describing motion based on graphs can be found in Chapter 18.

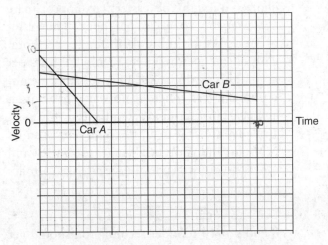

1. The velocity-time graphs represent the motion of two cars, Car A and Car B. Justify all answers thoroughly.

   (a) Which car is moving faster at the start of the motion?

   (b) Which car ends up farther from its starting point?

   (c) Which car experiences a greater magnitude of acceleration?

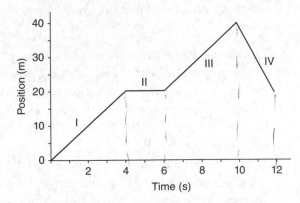

2. The following questions refer to the preceding position-time graph, which is the readout of an eastward-pointing motion detector. Justify all answers thoroughly.

   (a) Rank the speed of the object in each of the four labeled regions of the graph, from fastest to slowest. If the object has the same speed in two or more regions, indicate so in your ranking.

(b) What total distance did the object travel in the 12 s, including all parts of the motion?

(c) How far from the object's starting point did the object end up after the 12 s?

(d) Which of the following objects could reasonably perform this motion?
   (A) A baby crawling
   (B) A sprinter
   (C) A car on the freeway
   (D) A jet airplane during takeoff
   (E) An amoeba in a petri dish

3. A ball is dropped from rest near Earth. Neglect air resistance.[11] Justify all answers thoroughly.

   (a) About how far will the ball fall in 3 s?

   (b) The same ball is dropped from rest by an astronaut on the Moon, where the free-fall acceleration is one-sixth that on Earth. In 3 s, will the ball on the moon fall:
   (A) One-sixth as far as the ball on Earth
   (B) One-36th as far as the ball on Earth
   (C) The same distance as the ball on Earth
   (D) Six times as far as the ball on Earth
   (E) Thirty-six times as far as the ball on Earth

Data Table

Projectile	Initial Horizontal Speed (m/s)	Initial Vertical Speed (m/s)	Time of Flight (s)
A	40.0	29.4	6.00
B	60.0	19.6	4.00
C	50.0	24.5	5.00
D	80.0	19.6	4.00

4. Four projectiles, A, B, C, and D, were launched from and returned to level ground. The preceding data table shows the initial horizontal speed, initial vertical speed, and time of flight for each projectile. Justify all answers thoroughly.

   (a) Rank the projectiles by the horizontal distance traveled while in the air.

   (b) Rank the projectiles by the maximum vertical height reached.

   (c) Rank the projectiles by the magnitude of their acceleration while in the air.

---

[11]*Always* neglect air resistance, unless it is extremely, abundantly, and unambiguously clear from the problem's context that air resistance is important (i.e., talking about "terminal velocity").

› Solutions to Practice Problems

1. (a) On a velocity-time graph, speed is read off of the vertical axis. At time = 0, Car A has a higher vertical axis reading than Car B, so Car A is moving faster.

   (b) Displacement is determined by the area under a velocity-time graph. Car A's graph is a small triangle; Car B's graph is a trapezoid of obviously larger area. Both cars' graphs are always above the horizontal axis, so both cars move in the same direction the whole time; Car B moves farther away from its starting point.

   (c) Acceleration is the slope of the velocity-time graph. Car A's graph is steeper, so its acceleration is larger. [Sure, the slope of Car A's graph is negative, but that just means acceleration is in the negative direction, whatever that is; the question asks for the "magnitude" of the acceleration, meaning the amount, regardless of direction.]

2. (a) IV > I = III > II. Speed is the steepness on a position-time graph, without reference to direction (i.e., whether the slope is positive or negative). Segment IV is steepest. Segments I and III seem to be the same steepness. Segment II has 0 slope, so it represents an object that doesn't move.

   (b) It's a position-time graph, so read the vertical axis to figure out where the object is at any time. The object travels from its original position at $x = 0$ m to $x = 40$ m, then backtracks another 20 m. The total distance traveled is 60 m.

   (c) It's a position-time graph, so read the vertical axis to figure out where the object is at the beginning and after 12 s. At the beginning the object was at $x = 0$ m; after 12 s, the object was at position $x = 20$ m. The object traveled 20 m. (If your justification didn't explicitly mention that the object started at $x = 0$ m, or that you must find the *difference* between the final and initial positions, then it's incomplete.)

   (d) At its top speed in segment IV, the object travels 20 m in about 2 s. That's a speed of 10 m/s. If you're familiar with track and field, you'll know that the best sprinters run the 100-m dash in somewhere in the neighborhood of 10 s, so the sprinter is an obvious choice. It might be easier[12] to approximate a conversion to miles per hour. The result of 1 m/s is a bit more than 2 miles per hour. This object goes between 20 and 25 miles per hour. This is the speed of a car on a neighborhood street. There is no way a baby or an amoeba can keep up; takeoff speeds for most airplanes are at least in the high tens of miles per hour; and you'd be a danger to yourself and others if you drove on the freeway at 25 miles per hour.

3. (a) Use the equation $\Delta x = v_0 t + \frac{1}{2} a t^2$ with $v_0 = 0$ and $a = 10$ m/s per second. You should get about 45 m.

   (b) In the equation we use in (a), the time of 3 s is still the same, as is v_0. The only difference is the acceleration a, which is in the numerator and is neither squared nor square rooted. Therefore, reducing a by one-sixth also reduces the distance fallen by one-sixth. That's choice A. (By the way, if your answer is A but your justification included "setting up a proportion" or anything without specific reference to this equation, your answer is incorrect.)

4. (a) Horizontal speed remains constant throughout a projectile's flight. Use $\Delta x = v_0 t + \frac{1}{2} a t^2$ horizontally with the acceleration term equal to zero. That means you're multiplying the horizontal speed by the time of flight. This gives D > C > A = B.

   (b) Regardless of the time of flight, the vertical speed is directly related to the maximum height reached. Why? Use $v_f^2 = v_0^2 + 2a\Delta x$ vertically with $v_f = 0$ and $a = 10$ m/s per second. The bigger the v_0, the bigger the Δx. So A > C > B = D.

   (c) Easy—all objects in free fall have a downward acceleration of 10 m/s per second. A = B = C = D.

---

[12]For an American, who doesn't usually know from meters per second (m/s), anyway. If you'd prefer km/hr, multiply speeds in m/s by about 4 to get km/hr.

❯ Rapid Review

- In a position-time graph, the object's position is read from the vertical axis.

- In a position-time graph, the object's speed is the slope of the graph. The steeper the slope, the faster the object moves. If the slope is a front slash (/), the movement is in the positive direction; if the slope is a backslash (\), the movement is in the negative direction.

- In a velocity-time graph, the object's speed is read from the vertical axis. The direction of motion is indicated by the sign on the vertical axis.

- In a velocity-time graph, the object's acceleration is the slope of the graph.

- In a velocity-time graph, the object's displacement is given by the area between the graph and the horizontal axis. The location of the object can't be determined from a velocity-time graph; only how far it ended up from its starting point can be determined.

- In any case of accelerated motion when three of the five principal motion variables are known, the remaining variables can be solved for using the kinematic equations.

- To calculate the missing values in a motion chart, use the three kinematic equations listed below. Choose whichever equation works mathematically.

  (1) $v_f = v_0 + at$

  (2) $\Delta x = v_0 t + \dfrac{1}{2} at^2$

  (3) $v_f^2 = v_0^2 + 2a\Delta x$

When an object is in free fall, its acceleration is 10 m/s per second[10] toward the ground. "Free fall" means no forces other than the object's weight are acting on the object.

A projectile has no horizontal acceleration, and so it moves at constant speed horizontally. A projectile is in free fall, so its vertical acceleration is 10 m/s per second downward.

To find the vertical component of a velocity at an angle, multiply the speed by the sine of the angle. To find the horizontal component of a velocity at an angle, multiply the speed by the cosine of the angle. This always works, as long as the angle is measured from the horizontal.

The horizontal and vertical motion charts for a projectile must use the same value for time.

CHAPTER 11

Forces and Newton's Laws

IN THIS CHAPTER

Summary: A force is a push or a pull applied by one object on another object. This chapter describes the construction and use of free-body diagrams, which are key to approaching problems involving forces.

Definitions

- ✪ A **force** is a push or a pull applied by one object and experienced by another object.
- ✪ The **net force** on an object is the single force that could replace all the individual forces acting on an object and produce the same effect. Forces acting in the same direction add together to determine the net force; forces acting in opposite directions subtract to determine the net force.
- ✪ **Weight** is the force of a planet on an object near that planet.
- ✪ The **force of friction** is the force of a surface on an object. The friction force acts parallel to the surface. **Kinetic friction** is the friction force when something is moving along the surface and acts opposite the direction of motion. **Static friction** is the friction force between two surfaces that aren't moving relative to one another.
- ✪ The **normal force** is also the force of a surface on an object. The normal force acts perpendicular to the surface.
- ✪ The **coefficient of friction** is a number that tells how sticky two surfaces are.
- ✪ **Newton's third law** says that the force of Object A on Object B is equal in amount and opposite in direction to the force of Object B on Object A.
- ✪ **Newton's second law** states that an object's acceleration is the net force it experiences divided by its mass, and is in the direction of the net force.

Describing Forces: Free-Body Diagrams

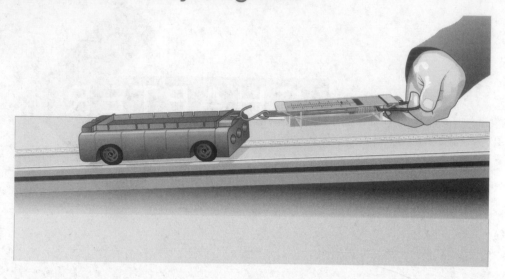

A force is a push or a pull applied by one object and experienced by another object. A force in the laboratory is often measured by a spring scale, as in the preceding picture. In AP Physics 1 we have to understand two aspects of forces. First, we have to describe the application of the force: What are the objects involved, and how much force is applied and in which direction? Next, we have to connect the net force acting on an object to that object's change in velocity.

Start with correct language: an object can "experience" a force, but an object cannot "have" a force. Don't let yourself say, "Ball A has a bigger force than Ball B"—that means nothing. "The net force on Ball A is bigger than on Ball B" is fine, as is "The Earth pulls harder on Ball A than on Ball B."

The canonical method of describing forces acting on an object is to draw a free-body diagram. A free-body diagram should include two elements:

(1) A labeled arrow representing each force, with each arrow beginning on the object and pointing in the direction in which the force acts
(2) A list of all the forces acting on the object, indicating the object applying the force and the object experiencing the force

On the AP Exam you'll be asked something like, "Draw and label the forces (not components) that act on the car as it slows down." This means "Draw a free-body diagram."

FACT: Only gravitational and electrical forces can act on an object without contact.[1]

Example 1: A car moving to the right on the freeway applies the brakes and skids to a stop.

Pretty much always start with the force of the Earth on the object, which is commonly known as its weight. Don't call this force "gravity"—that's an ambiguous term. Weight acts downward and doesn't require any contact with the Earth in order to exist.[2] Draw a downward arrow on the dot, label it "weight," and in the list write "Weight: force of Earth on the car."[3]

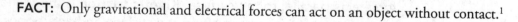

[1]In AP Physics 1, anyway.

[2]Chapter 15, on gravitation, explains more about how to find the weight of an object in a gravitational field.

[3]Why not call it "force of gravity on the car?" Well, because all forces must be exerted by an *object* on another object. Since when is "gravity" an object? ☺

Any other forces must be a result of contact with the car. What's the car touching? It's touching just the ground. Since the car is touching the ground, the ground exerts a normal force perpendicular to the surface. Draw an upward arrow on the dot, label it something like "F_n," and in the list write "F_n: The force of the ground on the car."[4]

Since the car is sliding along the ground, the ground exerts a force of (kinetic) friction. By definition, kinetic friction must always act in the opposite direction of motion—the car skids right along the ground, the friction force acts to the left. Draw a leftward arrow on the dot, label it "F_f," and in the list write "F_f: The force of the ground on the car."[5]

The car is not in contact with anything else, so we're done.

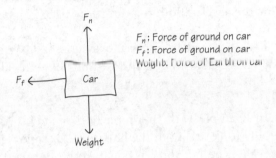

Whoa there! The car is moving to the right, so what about the force of its motion?

There's no such thing as the "force of motion." All forces must be exerted by an identifiable object; and all nongravitational and nonelectrical forces must be a result of contact. The car is not in contact with anything that pushes the car forward.

Then how is the car moving to the right?

It just is. It is critically important to focus *only* on the problem as stated. Questions about what happened before the problem started are irrelevant. Perhaps at first the car was pushed by the engine, or pulled by a team of donkeys, to start it moving; perhaps it had been in motion since the beginning of time. It doesn't matter. All that matters is that when we tune in to the action, the car is moving right and slowing down.

If the car had been pulled by a team of donkeys to start it moving, wouldn't we put the force of the donkeys on the car on the free-body diagram?

No, because the free-body diagram includes only forces that act *now*, not forces that acted earlier, or forces that will act in the future. If donkeys pulled the car, the force of the donkeys would appear on the free-body while the donkeys were actually pulling. After they let go and the car is slowing down, the donkeys might as well have never existed.

While it's important to learn how to draw a free-body diagram, it's just as important to learn how to *stop* drawing a free-body diagram. Don't make up forces. Unless you can clearly identify the source of the force, don't include the force.

---

[4] It makes no difference what you label the arrow, as long as you define the label in a list. You want to call it N instead of F_n? Be my guest.

[5] Yes, both the friction force and the normal force are properly listed as the force of the ground on the car.

Exam Tip from an AP Physics Veteran
If you see a problem involving forces, try drawing a free-body diagram for each object in the problem, or for a system including multiple objects. A free-body diagram will always be useful, even if you're not explicitly asked to make one.

Determining the Net Force

To determine the net force on an object, treat each direction separately. Add forces that point in the same direction; subtract forces that point in opposite directions. Or, if you know the acceleration in a direction, use $F_{net} = ma$.

FACT: When an object moves along a surface, the acceleration in a direction perpendicular to that surface must be zero. Therefore, the net force perpendicular to the surface is also zero.

In Example 1, the net force horizontally is equal to the force of friction, because that's the only force acting in the horizontal direction—there's no other force to add or subtract. Vertically, the net force is equal to the normal force minus the weight. But since the car is moving along the surface, the vertical acceleration and the vertical net force on the car are zero.

We can conclude, then, that the normal force on the car is equal to the car's weight. This isn't a general fact, though—a normal force is *not* always equal to an object's weight. If more vertical forces are acting, or if the surface is changing speed vertically (as in an elevator), the normal force can be different from the weight.

FACT: The kinetic friction force is equal to the coefficient of kinetic friction times the normal force.

$$F_f = \mu_k F_n$$

A good AP question might describe a second car, identical in mass and initial speed to the car in Example 1, but on a wet freeway. The question might ask you to explain why this second car skids to a stop over a longer distance.

The coefficient of friction is a property of the surfaces in contact. Here, since a wet road is less "sticky" than a dry road, the coefficient of friction has decreased. But since the second car is identical to the first, its weight and thus the normal force of the surface on the car is the same as before. Therefore, by the equation $F_f = \mu F_n$, the new car experiences a smaller force of friction.

With a smaller net force on the second car, its acceleration is also smaller by $F_{net} = ma$. Then the distance traveled during the skid depends on the car's acceleration by the kinematics equation (3), $v_f^2 = v_0^2 + 2a\Delta x$. Take the final speed v_f to zero and solve for Δx to see that acceleration a is in the denominator of the equation. Thus, a smaller acceleration means a larger distance to stop.

Static and Kinetic Friction

You may have learned that the coefficient of friction takes two forms: **static** and **kinetic** friction. Use the coefficient of static friction if something is stationary, and the coefficient of kinetic friction if the object is moving. The equation for the force of friction is essentially the same in either case: $F_f = \mu F_N$.

The only strange part about static friction is that the coefficient of static friction is a *maximum* value. Think about this for a moment—if a book just sits on a table, it doesn't need any friction to stay in place. But that book won't slide if you apply a very small horizontal pushing force to it, so static friction can act on the book. To find the maximum coefficient of static friction, find out how much horizontal pushing force will just barely cause the book to move; then use $F_f = \mu F_N$.

Newton's Third Law

FACT: The force of Object A on Object B is equal in amount and opposite in direction to the force of Object B on Object A. These two forces, which act on different objects, are called Newton's third law companion forces.[6]

In Example 1, then, what's the Newton's third law companion force to the normal force? It's tempting to say, "Oh, the weight." After all, the weight is equal to the normal force and is opposite in direction to the normal force. But that's wrong.

To find the companion force, look at the description of the force in the free-body diagram, and reverse the objects applying and experiencing the force. The normal force is the force of the ground on the car, and that acts upward. Therefore, the third law companion force is the force of the car on the ground, acting downward.

FACT: If the net force has both a vertical and a horizontal component, use the Pythagorean theorem to determine the magnitude of the net force, and use the tangent function to determine the direction of the net force.

In most AP problems, though, the net force will be zero in one or both directions. In Example 1, the magnitude of the net force is equal to only the magnitude of the friction force, because the vertical forces must subtract to zero.

If this were an AP problem, chances are it would ask about the connection between the net force and the change in the object's speed. We'll revisit this example later.

Forces at Angles

A force at an angle is drawn on a free-body diagram just like any other force. But when you're ready to do any analysis on the free-body diagram, start by breaking the angled force into components.

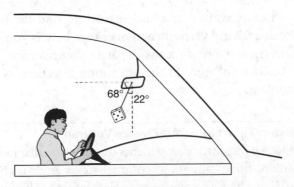

---

[6]Or, sometimes, this is called a Newton's third law force pair.

Example 2: A pair of fuzzy dice is hanging by a string from your rearview mirror, as shown in the preceding figure. You speed up from a stoplight. During the acceleration, the dice do not move vertically; the string makes an angle of $\theta = 22°$ with the vertical. The dice have mass 0.10 kg.

No matter what this problem ends up asking, you'll want to draw a free-body diagram. What forces act on the dice? Certainly the weight of the dice (the force of the Earth on the dice) acts downward. No electrical forces exist, so all other forces must be contact forces. The only object actually in contact with the dice is the string. The string pulls up on the dice at an angle, as shown in the picture. I labeled the force of the string on the dice "*T*," which stands for "tension"—which means the force of a string.

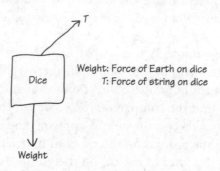

What about the force of the car on the dice?

What about it? The car is not in contact with the dice, the string is. It's the string, not the car, applying the force to the dice.

Determining the Net Force

It's likely that you'd be asked to determine the net force on these dice. But the tension acts both up *and* to the right. How do you deal with the vertical and horizontal forces, then?

FACT: When a force acts at an angle θ measured from the horizontal:

- The vertical component of that force is equal to the amount of the force itself times $\sin \theta$.
- The horizontal component of that force is equal to the amount of the force itself times $\cos \theta$.

Before you do any further work to find the net force, break all individual forces into horizontal and vertical components as best as you can. Here, the weight is already vertical. The tension becomes two separate components: $T\sin 68°$ goes in the vertical direction, and $T\cos 68°$ goes in the horizontal direction. Now you're ready to answer any possible problem.

Exam Tip from an AP Physics Veteran
Do *not* put force components on the same diagram as the force itself. You won't earn full credit. First, draw all forces at whatever angle is appropriate. Then, on a *separate diagram*, redraw the forces with the angled forces broken into components.

KEY IDEA

FACT: The acceleration of an object is F_{net}/m. This is the same thing as saying

$$F_{net} = ma$$

Whoa. I get that the net force is the horizontal $T\cos68°$, and that I can write that $T\cos68° = (0.1\ kg)(a)$. But the problem didn't give me an acceleration, it didn't give the tension—I'm stuck to solve for anything. The College Board screwed this problem up, right?

It's vanishingly unlikely that the problem is unsolvable as posed. You obviously can't ask questions of the College Board during the AP Exam. So if you're absolutely sure the exam is screwed up. you can just state where you think the problem is unclear, make up the information you need, and do your best. Chances are, though, that you need to find a creative alternate way to solve the problem.

Look at the problem statement: It said that the dice have a mass of 0.10 kg. This means that the weight of the dice is 1.0 N.[7] Since the dice are moving only in a horizontal direction, vertical acceleration (and the vertical net force on the dice) must be zero. Forces in opposite directions subtract to determine the net force. Here, that means that the up force must equal the down force of 1.0 N. The up force is $T\sin 68°$, which equals 1.0 N. Plug in from your calculator that the sine of 68 degrees is 0.93, and then solve to find the tension is 1.1 N.

Now deal with the horizontal direction. The horizontal net force is $T\cos 68°$, which is $(1.1\ N)(0.37) = 0.41\ N$. Since there's no vertical net force, 0.41 N to the right is the entire net force. And then $a = F_{net}/m$, so acceleration is 0.41 N/0.10 kg = 4.1 m/s per second. That's pretty much everything there is to calculate.

The Mistake

It's tempting to use this equation in all sorts of circumstances. For example, I'm sitting in a chair. Since I'm near the Earth, the force of the Earth on me is equal to my weight of 930 N. I know my mass is 93 kg. Use $a = F_{net}/m$. My acceleration must be 930 N/93 kg, or 10 m/s per second.

Um, no. I'm sitting in a chair. My speed isn't changing, so my acceleration is zero. The value of 10 m/s per second of acceleration means I'm in free fall. What went wrong?

I experience more forces than just the force of the Earth, of course. The chair is pushing up on me. Since I know that my acceleration is zero, the chair pushes up on me with 930 N of force. Now the up force and the down force on me subtract to zero. Phew.

Only the *net* force equals mass times acceleration. Never set a force equal to *ma* unless it's the net force.

What Else Could You Be Asked?

The AP Exam doesn't like to ask for calculations. So what else could be asked relating to Example 2?

---

[7] On the Earth, 1 kg of mass weighs 10 N. This fact is discussed in more detail in Chapter 15, Gravitation.

Here's one thought: If the dice were to instead hang from a bigger angle than 22° from the vertical, would the tension go up, go down, or stay the same?

The best way to answer this type of question is to make the calculation, and then explain what part of the calculation leads to the correct answer. That's the whole method behind answering questions that involve qualitative-quantitative translation, as discussed in Chapter 8.

Make up a bigger angle: call it 60° from the vertical. (Or choose any number; just make a significant difference in the new situation. Don't choose 23°.) Start back from the beginning: The dice still have a 0.10-kg mass, and the weight of the dice is still 1.0 N. The vertical acceleration is still zero, which means we can set the up force equal to the down force. But now the up force has changed, from $T \sin 68°$ to $T \sin 30°$.[8] Now we set $T \sin 30°$ equal to 1.0 N, giving a tension of 2.0 N.

The answer, then, is that the tension increases. The weight remains the same and the vertical component of tension must stay the same, but since to calculate tension we end up dividing the 1.0-N weight by the sine of the angle from the horizontal, a smaller angle from the horizontal gives a bigger tension.

Exam Tip from an AP Physics Veteran

If you are asked whether something increases, decreases, or stays the same, you might want to start by making a calculation to see numerically what happens to the answer. Be sure to explain *why* the calculation came out the way it did.

Inclined Planes

Treat objects on inclines the same as any other objects. Draw a free-body diagram, break angled forces into components, and use $a = F_{net}/m$ in each direction. The only major difference is that you don't use horizontal and vertical components for the forces. Instead, you look separately at the forces parallel to the incline and at the forces perpendicular to the incline.

Any normal force will be perpendicular to the incline, and so won't have to be broken into components as long as the object is moving up or down the incline. Any friction force will be parallel to the incline and so won't have to be broken into components. It's the weight—the force of the Earth—that will be broken into components.

FACT: On an incline of angle θ (measured from the horizontal), break the weight into components:

- The component of the weight that is parallel to the incline is equal to the weight times sin θ.
- The component of the weight that is perpendicular to the incline is equal to the weight times cos θ.

Example problems and extra drills on this frequently tested topic are available in Chapter 18.

---

[8]Remember we had to measure from the horizontal according to the fact on the previous page.

Multiple Objects

When two masses are connected over a pulley, it's often easiest to start by considering both objects as a single system. Draw the free-body diagram for the entire system, and use $a = F_{net}/m$ to find the acceleration of the system. Then, if you need to find the tension in the connecting rope, or if you need to talk about just one of the two connected objects, draw a new free-body diagram just for that object.

FACT: One rope has just one tension.[9]

An alternative approach is to start by drawing two separate free-body diagrams, one for each object. Write $F_{net} = ma$ for each object separately. Then, recognizing that the tension is the same in each equation, solve algebraically for the acceleration and tension.

› Practice Problems

Note: Extra drills on problems including ropes and inclined planes can be found in Chapter 18.

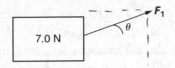

7.0 N

1. A 7.0-N block sits on a rough surface. It is being pulled by a force $\mathbf{F_1}$ at an angle $\theta = 30°$ above the horizontal, as shown above. The block is initially moving to the right with speed 5 m/s. The coefficient of friction between the block and the surface is $\mu = 0.20$. Justify all answers.

(a) Is it possible for the block to be slowing down? If so, give a possible value of the magnitude of F_1 that would allow the block to slow down. If not, explain why not with reference to Newton's second law.

(b) In order to double the block's initial speed to 10 m/s, how must the magnitude of the force F_1 change?
(A) It must double.
(B) It must quadruple.
(C) It does not have to change.

2. A drag-racing car speeds up from rest to 22 m/s in 2 s. The car has mass 800 kg; the driver has mass 80 kg.

(a) Calculate the acceleration of the drag racer.
(b) Calculate the net force on the drag racer.

(c) Which experiences a greater net force?
(A) The driver
(B) The car
(C) Both the driver and the car experience the same net force.

3. A car slides up a frictionless inclined plane. How does the normal force of the incline on the car compare with the weight of the car?

(A) The normal force must be equal to the car's weight.
(B) The normal force must be less than the car's weight.
(C) The normal force must be greater than the car's weight.
(D) The normal force must be zero.

4. Bert, Ernie, and Oscar are discussing the gas mileage of cars. Specifically, they are wondering whether a car gets better mileage on a city street or on a freeway. All agree (correctly) that the gas mileage of a car depends on the force that is produced by the car's engine—the car gets fewer miles per gallon if the engine must produce more force. Whose explanation is completely correct?

Bert says: Gas mileage is better on the freeway. In town the car is always speeding up and slowing down because of the traffic lights, so because $F_{net} = ma$ and acceleration is large, the engine must produce a lot of force. However, on the freeway, the

---

[9]This is true unless the rope is tied to or connected over a mass. For example, if the pulley itself had mass, then the rope can have different tensions on each side of the pulley. But that's a rare happening, and that certainly shouldn't require any calculation.

car moves with constant velocity, and acceleration is zero. So the engine produces no force, allowing for better gas mileage.

Ernie says: Gas mileage is better in town. In town, the speed of the car is slower than the speed on the freeway. Acceleration is velocity divided by time, so the acceleration in town is smaller. Because $F_{net} = ma$, then, the force of the engine is smaller in town giving better gas mileage.

Oscar says: Gas mileage is better on the freeway. The force of the engine only has to be enough to equal the force of air resistance—the engine doesn't have to accelerate the car because the car maintains a constant speed. Whereas in town, the force of the engine must often be greater than the force of friction and air resistance in order to let the car speed up.

› Solutions to Practice Problems

1. (a) When an object slows down, its acceleration (and therefore the net force it experiences) is opposite the direction of its motion. Here, the motion is to the right, so if the net force is left, it will slow down. To make the net force to the left, choose a value of F_1 such that the rightward component $F_1\cos(30°)$ is less than the friction force acting left.

   Choosing a value is a bit tricky: the value of the friction force itself depends on F_1, because the normal force on the block is 7 N minus the vertical component of F_1. Try choosing an F_1 much smaller than the block's weight, like 1 N. Then the normal force on the block is (7 N) – (1 N)sin 30° = 6.5 N. The friction force becomes $\mu F_n = 1.3$ N. The component of F_1 pulling right is 0.9 N, so the net force will be to the left as required.

   (Any value of F_1 that's less than a bit over 1.4 N will work here. Try it.)

   (b) While the net force is related to acceleration, the net force has no effect on an object's speed. Beyond that, no one has said anything about what happens before the problem, about how that initial speed came about. The forces can all be as indicated, and the object can have any initial speed.

2. (a) The car's speed changes by 22 m/s in 2 s. So the car changes its speed by 11 m/s in 1 s, which is what is meant by an acceleration of 11 m/s per second.

   (b) Newton's second law says that the net force on the racer is the drag racer's mass of 800 kg times the 11 m/s per second acceleration. That gives a net force of 8,800 N.

   (c) The driver and the car must experience the same acceleration because they move together; when the car changes its speed by 11 m/s in one second, so does the driver.[10] To calculate the net force on the driver, the driver's 80-kg mass must be used in Newton's second law, $F_{net} = ma$. With the same a and a smaller mass, the driver experiences a smaller net force (and the car experiences a greater net force).

3. (B) The normal force exerted on an object on an inclined plane equals mg (cos θ), where θ is the angle of the incline. If θ is greater than 0, then cos θ is less than 1, so the normal force is less than the object's weight.

4. Although Bert is right that acceleration is zero on the freeway, this means that the *net* force is zero; the engine still must produce a force to counteract air resistance. This is what Oscar says, so his answer is correct. Ernie's answer is way off—acceleration is not velocity/time, acceleration is a *change* in velocity over time.

---

[10]Otherwise, the driver would fall out of the car.

› Rapid Review

- Only gravitational and electrical forces can act on an object without contact (in AP Physics 1).

- When an object moves along a surface, the acceleration in a direction perpendicular to that surface must be zero. Therefore, the net force perpendicular to the surface is also zero.

- The friction force is equal to the coefficient of friction times the normal force, $F_f = \mu F_n$.

- The force of Object A on Object B is equal in amount and opposite in direction to the force of Object B on Object A. These two forces, which act on different objects, are called Newton's third law companion forces.

- If the net force has both a vertical and a horizontal component, use the Pythagorean theorem to determine the magnitude of the net force, and use the tangent function to determine the direction of the net force.

- When a force acts at an angle θ measured from the horizontal:

  - The vertical component of that force is equal to the amount of the force itself times $\sin \theta$.

  - The horizontal component of that force is equal to the amount of the force itself times $\cos \theta$.

- The acceleration of an object is F_{net}/m (which is the same thing as saying $F_{net} = ma$).

- On an incline of angle θ (measured from the horizontal), break the weight into components:

  - The component of the weight that is parallel to the incline is equal to the weight times $\sin \theta$.

  - The component of the weight that is perpendicular to the incline is equal to the weight times $\cos \theta$.

- One rope has just one tension.

CHAPTER 12

Collisions: Impulse and Momentum

IN THIS CHAPTER

Summary: Whenever you see a collision, the techniques of impulse and momentum are likely to be useful in describing or predicting the result of the collision. In particular, momentum is conserved in all collisions—this means that the total momentum of all objects is the same before and after the collision. When an object (or a system of objects) experiences a net force, the impulse momentum theorem $\Delta p = F \cdot \Delta t$ can be used for predictions and calculations.

Definitions

✪ A moving object's **momentum** is its mass times its velocity. Momentum is in the direction of motion.

✪ **Impulse** is defined as a force multiplied by the time during which that force acts. The net impulse on an object is equal to the change in that object's momentum.

✪ A **system** is made up of several objects that can be treated as a single thing. It's important to define the system you are considering before you treat a set of objects as a system.

✪ While total momentum is conserved in all collisions, kinetic energy is conserved only in an **elastic collision**.

Momentum is a useful quantity to calculate because it is often **conserved**; that is, the total amount of momentum available in most situations cannot change. Whenever you see a collision, the techniques of impulse and momentum are most likely to be useful. Try impulse and momentum first, before trying to use force or energy approaches.

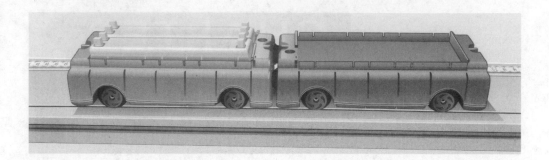

The Impulse-Momentum Theorem

Here is the impulse-momentum theorem:

$$\Delta p = F \cdot \Delta t$$

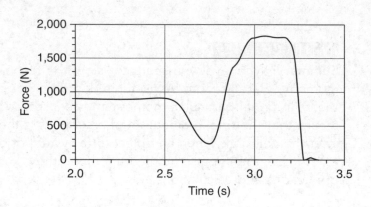

Example 1: A teacher whose weight is 900 N jumps vertically from rest while standing on a platform scale. The scale reading as a function of time is shown in the preceding figure.

A force versus time graph is essentially an invitation to calculate impulse. Since impulse is defined as $F \cdot \Delta t$, from a force versus time graph, impulse is the area under the graph.

Strategy: When you need to take the area of an experimental graph, approximate as best you can with rectangles and triangles.

This graph for Example 1 is tricky—impulse calculations should use the *net* force on an object. The scale reading on the vertical axis of the graph is not the net force; the net force is the scale reading in excess of the person's 900-N weight.

To estimate the impulse given to the jumper in this example, draw a horizontal line at the 900-N mark as a zero point for calculating the net force:

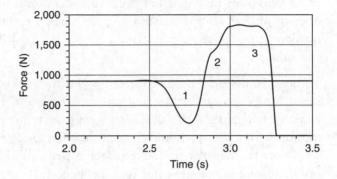

Regions 1 and 2 have somewhat close to the same area, one above and one above the zero net force line; so these areas cancel out. Region 3 looks somewhat like a rectangle, with base something like 0.25 s and height (1,800 N − 900 N) = 900 N. The impulse is then 900 N times 0.25 s, or something like 220 N·s.

Exam Tip from an AP Physics Veteran
You must be comfortable with this kind of rough approximation. Sure, the impulse could well be more like 228 N·s, or 210 N·s. Who cares? On a free-response item it will be the reasoning behind your calculation that earns credit, much more so than the answer itself. On a multiple-choice item, the choices might be far separated, like (A) 200 N·s; (B) 2,000 N·s; (C) 20,000 N·s; (D) 200,000 N·s. Isn't the choice obvious?

Impulse by itself doesn't say much. The more interesting question about this jumping teacher in Example 1 is the speed with which he leaves the scale. Since impulse is the change in an object's momentum, you know the teacher changed his momentum by 220 N·s. Since he started from rest, his momentum right after leaving the scale is also 220 N·s. Finally, momentum is mass times speed. The teacher's mass is 90 kg,[1] so plug in to the equation $p = mv$: (220 N·s) = (90 kg)v. His speed is 2.4 m/s, or thereabouts.

Conservation of Momentum

Example 2: Cart A, of mass 0.5 kg, moves to the right at a speed of 60 cm/s. Cart B, of mass 1.0 kg, is at rest. The carts collide.

FACT: In any system in which the only forces acting are between objects in that system, momentum is conserved. This effectively means that momentum is conserved in *all* collisions.

---

[1]This is because his weight is 900 N, and on Earth 1 kg weighs 10 N.

Define the system for momentum conservation in Example 2—just the two carts. They apply a force to each other in the collision, but that's it, so momentum is conserved.[2]

A common task in a problem with a collision involves calculating the speed of one or both objects after the collision. Even when a collision-between-two-objects question is qualitative or conceptual in nature, it's often a good idea to try calculating speeds after a collision.

To do this, define a positive direction and then make a chart indicating the mass m and speed v of each cart before and after the collision. I use a "prime" mark (′) to indicate when we're dealing with values after a collision rather than before. Indicate the direction of motion with a plus or minus sign on the velocities.

(Note that it's okay to use centimeters per second [cm/s] rather than meters per second [m/s], as long as you are consistent throughout.)

$$m_A = 0.5 \text{ kg}$$
$$m_B = 1.0 \text{ kg}$$
$$v_A = +60 \text{ cm/s}$$
$$v_B = 0$$
$$v_A' = ?$$
$$v_B' = ?$$

Then write the equation for conservation of momentum.

Uh, where do you get such an equation? I know it's not on the equation sheet on the old or new AP physics exams.

The relevant equation comes from the definition of "conservation," meaning an unchanging quantity. The total change in momentum for the system of the two carts must be zero. Any momentum lost by Cart A is gained by Cart B. Set zero equal to Cart A's change in momentum, plus Cart B's change in momentum:

$$0 = (p_A' - p_A) + (p_B' - p_B)$$

Then, knowing that $p = mv$, plug in what you know. I'm going to leave off the units to make the mathematics clearer; since the table above has values and units, it's clear what units are intended.

$$0 = [(0.5) \, v_A' - (0.5)(60)] + [(1.0)v_B' - 0]$$

Is this solvable? Not yet, because it's only one equation with two variables. The information about this collision is incomplete. The collision in Example 2 could thus have all sorts of results.

One possibility is that the carts stick together. In that case, the carts share the same speed: in the notation above, $v_A' = v_B'$. That makes the calculation solvable; replace the vs with a single variable to get $v = 20$ cm/s.

Perhaps the problem statement continues to tell us the speed of one of the carts after the collision. Then the problem is solvable: plug in the value given, and solve for the other v.

---

[2]If friction between the track and the carts were significant, then sure, momentum wouldn't be conserved—the track wasn't considered part of the system, and it's applying a force to the carts. But *even with friction*, if you consider the moments just before and just after the collision, momentum will be essentially conserved. See the following discussion for a time when the kmomentum of a system is *not* conserved.

The only tricky part here would be, say, if Cart A rebounded after the collision. Then v_A' would take a negative value. But the solution would be approached the same way.

When Is the Momentum of a System *Not* Conserved?

The simple answer goes back to the definition of momentum conservation: The momentum of a system is *not* conserved when a force is exerted by an object that's not in the system.

> **Example 3:** Two identical balls are dropped from the same height above the ground, such that they are traveling 50 cm/s just before they hit the ground. Ball A rebounds with speed 50 cm/s; Ball B rebounds with a speed of 10 cm/s. Each is in contact with the ground for the same amount of time.

Define the system here. If the system is just Ball A, say, then is the momentum of Ball A conserved? Of course not! The problem says that Ball A rebounds, which means it changed its direction and thus its momentum.

Mistake: It's tempting to say that since Ball A didn't change its mass, and since its speed was 50 cm/s before *and* after the collision, that Ball A didn't change its momentum. This is not correct; momentum has direction. An object that changes direction loses all its momentum and then gains some more. If Ball A had mass 2 kg, then it lost 1 N·s of momentum in stopping, and then gained another 1 N·s of momentum in order to rebound—for a total change in momentum of 2 N·s.

> *Wait.* You said in the preceding *Fact* that momentum is conserved in all collisions. What happened?

Well, yes, momentum *is* conserved in all collisions, if you define the system to include the two (or three) objects that are colliding. In Example 3, Ball A is effectively colliding with the entire Earth. If we consider the system of the Earth and Ball A, then momentum is, in fact, conserved. The change in Ball A's momentum is equal to the change in the Earth's momentum. Since the Earth is so mind-bogglingly massive, its speed won't change in any measureable amount.

Mistake: The total momentum for a system of objects is always the same. So in a single collision, the total momentum cannot change. In a problem like Example 3, though, Balls A and B are involved in two separate collisions. Therefore, they can't be part of the same system! Don't use "conservation of momentum" as a reason for anything about Balls A and B to be equal, when Balls A and B are involved in separate collisions.

The point is that momentum conservation is not an effective approach to consider when a ball collides with the entire Earth.[3] Instead, use the impulse-momentum theorem to find out what you can.

The easy question is as follows: Which ball changes its momentum by a greater amount? That'd be Ball A. Both balls lost the same amount of momentum in coming to a brief rest, then rebounded; since Ball A rebounded faster, and since the balls have the same mass, Ball A changed its momentum by a greater amount.[4]

---

[3]Or equivalently, this is not an effective approach to consider when a car collides with a concrete pillar, or a bird collides with the window of a building, etc.

[4]If you need to make up a mass of 2 kg for each ball and plug in numbers (including a plus and minus sign for the direction of velocity) to calculate the total momentum change for each ball, feel free. That's not a bad approach if the words are confusing you.

The harder question is this: Which ball exerted a larger force on the ground during its collision? We know that momentum change equals force times time.[5] With the same time of collision, the bigger force is exerted by the ball with the greater momentum change—that's Ball A.

Similar reasoning can explain why airbags make a car safer. You[6] lose all your momentum in a crash regardless of how you come to rest. Airbags extend the time of the collision between you and the car. In the equation $\Delta p = F \cdot \Delta t$ with the same Δp, a bigger Δt gives a smaller F, so the force you experience is less in an airbag collision.

Elastic/Inelastic Collisions

In elastic collisions, the total kinetic energy of both objects combined is the same before and after the collision. A typical AP problem might pose a standard collision problem and then ask, "Is the collision elastic?" To figure that out, add up the kinetic energies ($\frac{1}{2}mv^2$) of both objects before the collision, add up the kinetic energies of both objects after the collision, and compare. If these kinetic energies are essentially the same, the collision is elastic. If the final kinetic energy is less than the initial kinetic energy, the collision was *not* elastic—kinetic energy was converted, generally to work done by nonconservative forces exerted by one colliding object on the other.[7]

> **Example 4:** Two carts of equal mass move toward each other with identical speeds of 30 cm/s. After colliding, the carts bounce off each other, each regaining 30 cm/s of speed, but now moving in the opposite direction.

Mistake: Never start a collision problem writing anything about kinetic energy. Always start with conservation of momentum. Only move on to kinetic energy conservation if you have to, that is, if you don't have enough information to solve with just momentum conservation, *and* if the problem is explicit in saying that the collision is elastic.

Is momentum conserved in this collision? Yes, and you don't have to do any calculations to show it. In a collision, momentum is always conserved because the only forces acting on the carts are exerted by the carts themselves.

Is kinetic energy conserved in this collision? You've got to do the calculation to check.

But the carts bounced off each other. Doesn't that automatically mean the collision is elastic?

No. When carts stick together, the collision cannot be elastic. But when carts bounce off each other, the collision might be elastic, or might not be.

Start with the kinetic energy before the collision. In this case, make up a mass for each cart: they're identical, so call them 1 kg each. Each cart has speed 0.30 m/s, so the kinetic energy of each cart before the collision is $\frac{1}{2}(1 \text{ kg})(0.30 \text{ m/s})^2 = 0.045$ J. The combined kinetic energy before collision is thus 0.090 J.

---

[5]**Mistake**: The relevant time in the impulse-momentum theorem is always the time of the collision, *not* the time it takes for a ball to fall through the air.

[6]Hopefully not *you*, personally.

[7]If the final kinetic energy is *greater* than the initial kinetic energy, something weird happened; like a coiled spring was released during the collision, or a firecracker exploded. You'll most often see this sort of "superelastic" collision when the objects are initially at rest and then they are blown apart.

Whoa there—One cart was moving right, the other left; that means the kinetic energies subtract, giving zero total kinetic energy. Right?

Wrong. Kinetic energy is a scalar, which means it has no direction. Kinetic energy can never take on a negative value. Always add the kinetic energies of each object in a system to get the total kinetic energy. After the collision, the calculation is the same: total kinetic energy is still 0.090 J. So the collision is, in fact, elastic.

2-d Collisions

Example 5: Maggie, of mass 50 kg, glides to the right on a frictionless frozen pond with a speed of 2.5 m/s. She collides with a 20-kg penguin. After the collision, the directions of the penguin's and Maggie's motion is shown in the following figure.

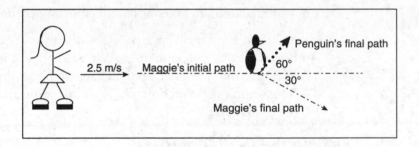

Strategy: When objects move in both an *x*- and a *y*-direction after a collision, analyze the collision with momentum conservation *separately* in each direction.

You will not likely be asked to do quantitative analysis of a two-dimensional collision, but you do need to understand conceptually how momentum conservation works here. Be able to explain how you would carry out the analysis of momentum conservation in each direction and be able to answer simple qualitative questions.

For example, who has a greater magnitude of momentum in the *y*-direction after collision? Before the collision, there was no momentum in the *y*-direction. After the collision, the total *y*-momentum must also be zero. Since both the penguin and Maggie are moving in the *y*-direction, their momentums must be equal and opposite so as to subtract to zero. The answer is neither—both the penguin and Maggie have the same amount of *y*-momentum.

What about the *y*-component of their velocities? We've already established that they have the same *y*-momentum, which is equal to mass times *y*-velocity. Since Maggie has the bigger mass, she must have the smaller *y*-component of velocity.

Is momentum conserved in the *x*-direction? Of course it is. The total momentum before collision is all due to Maggie's movement: (50 kg)(2.5 m/s) = 125 N·s, all in the *x*-direction. After collision, the total *x*-momentum is also 125 N·s. The *x*-component of the penguin's momentum after collision is just his momentum *mv* times the cosine of 60°; Maggie's *x*-momentum is her momentum times the cosine of 30°. In this problem, the only way to get values for these components is to do some complicated algebra, which is beyond the scope of AP Physics 1. But you should be able to explain everything about this collision in words, as discussed here.

Motion of the Center of Mass

FACT: The center of mass of a system of objects obeys Newton's second law.

Two common examples illustrate this fact:

Example: Imagine that an astronaut on a spacewalk throws a rope around a small asteroid, and then pulls the asteroid toward him. Where will the asteroid and the astronaut collide?

Answer: at the center of mass. Since no forces acted except due to the astronaut and asteroid, the center of mass must have no acceleration. The center of mass started at rest, and stays at rest, all the way until the objects collide.

Example: A toy rocket is in projectile motion, so that it is on track to land 30 m from its launch point. While in the air, the rocket explodes into two identical pieces, one of which lands 35 m from the launch point. Where does the first piece land?

Answer: 25 m from the launch point. Since the only external force acting on the rocket is gravity, the center of mass must stay in projectile motion, and must land 30 m from the launch point. The two pieces are of equal mass, so if one is 5 m beyond the center of mass's landing point, the other piece must be 5 m short of that point.

Finding the Center of Mass

Usually the location of the center of mass (cm) is pretty obvious . . . the formal equation for the cm of several objects is

$$Mx_{cm} = m_1x_1 + m_2x_2 + \cdots$$

Multiply the mass of each object by its position, and divide by the total mass M, and voila, you have the position of the center of mass. What this tells you is that the cm of several equal-mass objects is right in between them; if one mass is heavier than the others, the cm is closer to the heavy mass.

› Practice Problems

1. A 2-kg coconut falls from the top of a tall tree, 30 m above a person's head. The coconut strikes and comes to rest on the person's head. Justify all answers thoroughly.

   (a) Calculate the magnitude momentum of the coconut just before it hits the person in the head.

   (b) Calculate the magnitude and direction of the impulse experienced by the coconut in colliding with the person's head.

   (c) The person's head experienced a force of 10,000 N in the collision. How long was the coconut in contact with the person's head?
   (A) Much more than 10 seconds
   (B) Just a bit more than one second
   (C) Just a bit less than one second
   (D) Much less than 1/10 second

   (d) In a different situation, explain how it could be possible for an identical coconut dropped from the same height to hit the person's head, but produce *less* than 10,000 N of force.

2. A car on a freeway collides with a mosquito, which was initially at rest. Justify all answers thoroughly.

(a) Did the total momentum of the car-mosquito system increase, decrease, or remain the same after the collision?

(b) Did the momentum of the mosquito increase, decrease, or remain the same after the collision?

(c) Did the momentum of the car increase, decrease, or remain the same after the collision?

(d) Which changed its speed by more in the collision, the car or the mosquito? (Or did they change speed by the same amount?)

(e) Which changed its momentum by more in the collision, the car or the mosquito? (Or did they change momentum by the same amount?)

(f) Which experienced a greater impulse in the collision, the car or the mosquito? (Or did they experience the same impulse?)

(g) Which experienced a greater magnitude of net force during the collision, the car or the mosquito? (Or did they experience the same net force?)

3. Car A has a mass of 1,500 kg and travels to the right with a speed of 20 m/s. Car B initially travels to the left with a speed of 10 m/s. After the vehicles collide, they stick together, moving left with a common speed of 5 m/s. Justify all answers thoroughly.

(a) Calculate the mass of Car B.

(b) This collision is not elastic. Explain why not.

(c) Describe specifically a collision between these two cars with the same initial conditions, but which is *not* elastic, and in which the cars bounce off one another.

(d) Is the collision elastic when Car B remains at rest after the collision?

› Solutions to Practice Problems

1. (a) Momentum is mass times speed. To find the coconut's speed, use kinematics with $v_0 = 0$, $a = 10$ m/s per second, and $\Delta x = 30$ m. The equation $v_f^2 = v_0^2 + 2a\Delta x$ solved for v_f gives 24 m/s. Multiplying by the 2-kg mass gives a momentum of 48 N·s.[8]

(b) Impulse is the change in momentum and is in the direction of the net force experienced by an object. The coconut's momentum after colliding with the head is zero—the coconut comes to rest. So its change in momentum, and thus the magnitude of the impulse it experiences, is 48 N·s. The direction of this impulse is upward, because the net force on the coconut must be opposite its speed in order to slow it down.

(c) Impulse is also equal to force times the time interval of collision. Setting 48 N·s equal to $(10,000 \text{ N})(\Delta t)$, we find the time interval of collision is 48/10,000 of a second—much less than 1/10 second, even without reference to a calculator.

(d) The impulse-momentum theorem says that $\Delta p = F\Delta t$. Solving for force, $F = \frac{\Delta p}{\Delta t}$. Here the momentum change has to be the same no matter what—the coconut will be traveling 24 m/s and will come to rest on the person's head. But if the person is wearing a soft helmet, or if the coconut has a rotten spot on it somewhere, then the time of collision could be larger than before. Since Δt is in the denominator of the force equation, a bigger time interval of collision leads to a smaller force on the coconut (and therefore on the person's head).

2. (a) Momentum is conserved when no forces are exerted, except for those on and by objects in the system. Here the only forces are of the car on mosquito and mosquito on car. Therefore, momentum was conserved. That means that the total momentum of the car-mosquito system remains the same.

---

[8]If you used units of kg·m/s, that's fine, too.

(b) The mosquito went from rest to moving free-way speeds after it hit the car. The mosquito's mass didn't change.[9] Momentum is mass times speed, so the mosquito's momentum increased.

(c) Since total momentum of the car-mosquito system doesn't change, and the mosquito gained momentum, the car has to lose that same amount of momentum.

(d) The mosquito's speed went from, say, zero to 60 miles per hour. While the car must lose the same amount of momentum that the mosquito gained, the car's mass is so much larger than the mosquito's that the car's speed will hardly change. And you knew that, because a car hitting a mosquito on the freeway doesn't cause the car to stop.

(e) The momentum change is the same for both, because total momentum remains unchanged. Any momentum gained by the mosquito must be lost by the car.

(f) Impulse is the same thing as momentum change, so the same for both.

(g) Newton's third law says the force of the mosquito on the car is equal to the force of the car on the mosquito. So they're equal.

3. (a) Before the collision, Car A has a momentum of 30,000 N·s to the right. If we call the mass of Car B "M_B," then Car B has momentum of M_B(10 m/s) to the left. Afterward, the total momentum is (M_B + 1,500 kg)(5 m/s) to the left. Let's call right the positive direction. Then the relevant equation for conservation of momentum is $30,000 - 10M_B = -5(1,500 + M_B)$, where I've left off the units so the algebra is clearer. Solve for M_B to get 7,500 kg. This makes sense—Car B was initially moving slower, yet after the collision the cars moved off together in the direction Car B was going. Car B must therefore have more momentum than Car A initially and more mass because it was going slower.

(b) "Elastic" means that kinetic energy (= $\frac{1}{2}mv^2$) of all objects combined is the same before and after collision. Before collision, Car A had 300 kJ of kinetic energy, and Car B had 375 kJ, for a total of 675 kJ before the collision.[10] After collision, the kinetic energy of the combined cars is 112 kJ. Kinetic energy was lost in the collision. (Note that it's legitimate to remember that collisions in which objects stick together can never be elastic.)

(c) Imagine that Car B keeps moving left, but much slower, say, 1 m/s. Momentum is conserved in a collision, regardless of whether the collision is elastic or not. The total momentum of Car A before collision is 30,000 N·s to the right; the total momentum of Car B before collision is 75,000 N·s to the left. This gives a total momentum of 45,000 N·s to the left before collision. If Car B moves 1 m/s after collision, it has 7,500 N·s of momentum to the left, leaving 37,500 N·s to the left for Car A. Dividing by Car A's 1,500-kg mass, Car A is found to be moving 25 m/s after the collision.

Now check total kinetic energy after collision. Car A has 469 kJ of kinetic energy and Car B has 4 kJ of kinetic energy, for a total of 473 kJ. Before the collision the total kinetic energy was 675 kJ, as calculated in (b). Therefore, kinetic energy is lost and the collision is inelastic. The whole point here is that not all collisions in which cars bounce are elastic.

(d) We need a total of 675 kJ afterward in order to have an elastic collision. Conservation of momentum means that the total momentum after collision is 45,000 N·s to the left. Since Car A is the only moving car, it has all that momentum. Dividing by Car A's 1,500-kg mass, we find Car A moving 30 m/s. Only Car A has kinetic energy, too; its kinetic energy is $\frac{1}{2}mv^2$ = 675 kJ, so the collision is elastic.

---

[9] . . . though its mass was likely redistributed around the windshield a bit.

[10] No, the total is not −75 kJ. Kinetic energy is a scalar, meaning it cannot have a direction; and kinetic energy cannot be negative. The total kinetic energy of a system is the sum of all the kinetic energies of the constituent objects, regardless of which way the objects are moving.

› Rapid Review

- In any system in which the only forces acting are between objects in that system, momentum is conserved. This effectively means that momentum is conserved in *all* collisions.

- The center of mass of a system of objects obeys Newton's second law.

- The impulse-momentum theorem is $\Delta p = F \cdot \Delta t$.

- The impulse-momentum theorem is always valid, but it is most useful when objects collide.

- The only time when momentum of a system is *not* conserved is when a force is exerted by an object that's not in the system.

CHAPTER 13

Work and Energy

IN THIS CHAPTER

Summary: An object possesses kinetic energy by moving. Interactions with other objects can create potential energy. Work is done when a force acts over a distance parallel to that force. When work is done on an object (or on a system of objects), kinetic energy can change. This chapter shows you how to recognize the different forms of energy and how to use them to make predictions about the behavior of objects.

Definitions

✪ **Kinetic Energy** is possessed by any moving object. It comes in two forms:

   1. _Translational Kinetic Energy_ is

$$\tfrac{1}{2}mv^2$$

   It exists when an object's center of mass is moving.

   2. _Rotational Kinetic Energy_ is

$$\tfrac{1}{2}I\omega^2$$

   It exists when an object rotates.

✪ **Gravitational potential energy** is energy stored in a gravitational field. Near a planet, the formula is

$$GPE = mgh$$

where h is the vertical height above a reference position. A long way from a planet, the formula is

$$GPE = -G\frac{M_1 M_2}{d}$$

where d is measured from the planet's center.

✪ **Elastic potential energy,** also known as spring potential energy, is energy stored by a spring, given by

$$SPE = \tfrac{1}{2}kx^2$$

✪ **Internal Energy** can refer to two somewhat different ideas. Both refer to the concept that multi-object systems can store energy depending on how the objects are arranged in the system.

  ✪ *Microscopic internal energy* is related to the temperature of the object. As the object warms up, energy can be stored by the vibrations of molecules.

  ✪ *Internal energy of a two-object system* is just another way of saying "potential energy."

✪ **Mechanical Energy** refers to the sum of potential and kinetic energies.

✪ **Work** is done when a force acts on something that moves a distance parallel to that force.

✪ **Power** is defined as energy used per second, or work done per second.

Energy

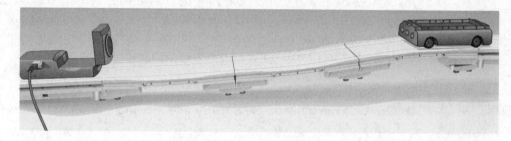

The cart pictured above has gravitational potential energy with respect to the location of the motion detector, because the cart is vertically higher than the detector. If the cart were released from rest, it would speed up toward the detector. It would be tempting to try to use the kinematics equations to determine the cart's maximum speed. However, since the track is curved, the cart's acceleration will be changing throughout its motion. Whenever acceleration changes, kinematics as studied in Chapter 10 are invalid. The methods of energy conservation, as described in this chapter, must be used.

The College Board's curriculum guide for AP Physics 1 makes much of the difference between **objects** and **systems**. A system is just a collection of objects. In your class, you might well have talked about an object's kinetic and potential energies. The thing is, the exam development committee doesn't like that language. Sure, an object can have kinetic energy, just by moving. But a single object technically cannot "have" potential energy.

Why not? Potential energy is always the result of an interaction between objects in a system. For example, gravitational potential energy (equation: mgh) exists only if an object is interacting with the Earth. The Earth-object system stores the potential energy, not just the object itself. Similarly, an object attached to a spring *cannot* store potential energy; the spring-object system stores the energy.

Look, I'm going to talk about objects "having" potential energy. It's okay with me if you talk about a block on a spring "having" potential energy because the spring is compressed. You'll still get pretty much everything right on the exam. Just know that the block only "has" potential energy because of its interaction with the spring; and that potential energy is sometimes referred to as the "internal energy of the block-spring system."

Work

FACT: Work is done when a force is exerted on an object[1] and that object moves parallel to the direction of the force.

The relevant equation is work = force times parallel displacement:[2]

$$\boxed{W = F\Delta x_{\parallel}}$$

When a force is exerted in the same direction as the object's motion, the work done is considered to be a positive quantity; when a force is exerted in the opposite direction of the object's motion, the work done is considered to be a negative quantity. The *net* work on an object is the algebraic sum of the work done by each force.

Example 1: A string applies a 10-N force to the right on a 2-kg box, dragging it at constant speed across the floor for a distance of 50 cm.

Let's calculate the work done by each force acting, and the net work done on the box. Start by sketching a free-body diagram for the box.

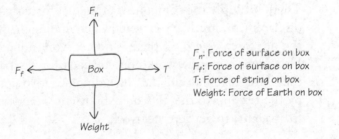

F_n: Force of surface on box
F_f: Force of surface on box
T: Force of string on box
Weight: Force of Earth on box

Strategy: Whenever you are calculating work done by a force with the equation $W = F\Delta x_{\parallel}$, always sketch the direction of the force and displacement vectors.

Consider each force separately. Start with the 10-N tension in the string. This force acts to the right. The displacement of the box is 50 cm (i.e., 0.50 m) to the right.

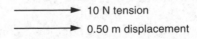

Since the force is in the same direction as the displacement, the work done by the tension is just 10 N times 0.50 m = +5 J.

Now consider the friction force. Since the box moves at constant speed, we know the left force (friction) equals the right force (tension); so the friction force has an amount of 10 N.

---

[1]Or on a system of objects—if a force is exerted on a system of objects and the system's center of mass moves parallel to the force, work was done.

[2]Or, equivalently, the equation is displacement times parallel force, which may sometimes be a more convenient expression.

Since the force is in the *opposite* direction of displacement, the work done by the friction force is 10 N times 0.50, with a negative sign. Thus, the work done by friction is –5 J. What about the force of gravity? The 20-N weight of the box points downward.

20 N weight   0.50 m displacement

Since no component of the weight is parallel to the displacement, the force of the Earth does zero work on the box.

Similarly, the normal force is straight upward while the displacement is to the right; since no component of the normal force is parallel to the displacement, the normal force does no work on the box.

> **Exam Tip from an AP Physics Veteran**
>
> If a force acts at an angle to the displacement, just break that force into components. The component perpendicular to the displacement does no work. The component parallel to the displacement can be multiplied by the displacement to get the work done.

Finding the Net Work

You can calculate the net work on the box in two ways:

1. First, determine the net force using a free-body diagram, like we showed in Chapter 11. Then, multiply the component of the net force that's parallel to the displacement by the displacement, just like you would when finding the work done by any force. In Example 1, the net force is zero; so there is no net work done on the box.
2. First, determine the work done by each force separately. Then, add the work done by each force algebraically (i.e., including negative signs). In Example 1, add the +5 J done by the tension to the –5 J done by friction (and the 0 J done by the normal force and the weight) to get zero net work.

Conservative Versus Nonconservative Forces

> **FACT:** A "conservative" force converts potential energy to other forms of mechanical energy when it does work. Thus, a conservative force does not change the mechanical energy of a system.

The amount of work done by a conservative force depends only on the starting and ending positions of the object, i.e., it's "path independent." The only conservative forces that you need to deal with on the AP Physics 1, Algebra-Based Exam are gravity and springs. When a spring does work on an object, energy is stored in the spring that can be recovered and converted back to kinetic energy. The sum of the potential and kinetic energy of the object-spring system is constant.

Conversely, a "nonconservative" force can change the mechanical energy of a system. Friction is the most common example: Work done by friction on an object becomes microscopic internal energy in the object, raising the object's temperature. That microscopic internal energy *cannot* be recovered and converted back to kinetic energy. Other nonconservative forces might include, for example, the propeller of an airplane—it does work on the airplane to increase the airplane's mechanical energy.

The Work-Energy Theorem

In your textbook, you'll see the work-energy theorem written as "net work = change in kinetic energy." That's certainly true—*net* work done on an object must change the object's kinetic energy. The tricky part is, net work must include work done by all forces, conservative and nonconservative.

I think it's easier to separate conservative and nonconservative forces. Work done by a nonconservative force (W_{NC}) changes the total mechanical energy of a system ($KE + PE$). I write the work-energy theorem as follows:

$$W_{NC} = (\Delta KE) + (\Delta PE)$$

Generally, the potential energy involved will be either that due to a spring, or due to a gravitational field. The kinetic energy includes *both* translational kinetic energy ($\frac{1}{2}mv^2$) and rotational kinetic energy[3] ($\frac{1}{2}I\omega^2$).

> **Example 2:** An archer pulls an arrow of mass 0.10 kg attached to a bowstring back 30 cm by exerting a force that increases uniformly with distance from 0 N to 200 N.

The AP Exam could ask all sorts of questions about this situation. Before you start doing any calculation, categorize the problem. There are only three ways to approach a mechanics problem: kinematics/Newton laws, momentum, and energy. There's no collision, so momentum is unlikely to be useful. The problem talks about a force; but that force is *changing*. A changing force means a changing acceleration, which means that kinematics equations are not valid. Only the work-energy theorem will be useful.

The only types of potential energy used in AP Physics 1 are due to gravity (mgh) and due to a spring ($\frac{1}{2}kx^2$). Which is involved here? The example says that the force of the string varies "uniformly," which means that the force gets bigger as the distance stretched gets bigger, just like a spring. So treat the bowstring just like a spring.

Mistake: An interesting question here might be "how much work does the archer do in pulling back the bowstring?" And you'd be tempted to use the definition of work, $W = F\Delta x_{\parallel}$. But, no, since the force of the archer on the string is changing, this equation for work does not apply. Instead, you must use the entire work-energy theorem.

To find the work done by the archer in pulling back the bowstring, write the work-energy theorem, considering the time from when he starts pulling until the maximum extension. Since the arrow is at rest before the archer starts pulling, and is *still* at rest when the string is pulled all the way back, the change in kinetic energy is zero. The potential energy of a spring is zero at the equilibrium position and is $\frac{1}{2}kx^2$ at full extension. The work done by the archer is a nonconservative force, since it changes the mechanical energy of the string-arrow system. We get

$$W_{NC} = (0) + (\frac{1}{2}kx^2 - 0)$$

But this isn't solvable yet—we know the distance x the archer pulled to be 0.30 m (i.e., 30 cm). But what is k, the spring constant of the "spring"? Use the equation $F = kx$ when the bow is fully extended. The problem says the maximum force the archer pulls with is 200 N when the string is extended 0.30. Plugging into $F = kx$ and solving for k gives $k = 670$ N/m.[4]

Now the work done by the archer is $\frac{1}{2}(670 \text{ N/m})(0.30 \text{ m})^2 = 30$ J.

---

[3] See Chapter 14 for further discussion of rotational kinetic energy.

[4] No, *not* 666.6666666666 N/m. Use two significant figures, unless you want your 10th-grade chemistry teacher to have heart palpitations.

The AP Exam could certainly ask for this calculation. Or, the exam might ask, "If the archer instead pulls back 60 cm, what will happen to the work done by the archer?" The work-energy theorem still applies, the kinetic energy terms still go away, and the work done is still $\frac{1}{2}kx^2$. The spring constant is a property of a spring (or in this case, a bowstring). So k doesn't change. We doubled the displacement from equilibrium, x. Since the variable x is squared, then we don't multiply the work done by a factor of two when we double x; we multiply the work done by a factor of 2^2 (i.e., by a factor of four). The archer has to do four times as much work.

How about finding out how fast the arrow would be traveling when the archer shoots it? We cannot use kinematics with a varying net force or a varying acceleration. Use the work-energy theorem again; except this time, let the problem start when the archer released the bowstring, and end when the string gets back to the equilibrium position and the arrow is released. Now, since no nonconservative forces act,[5] the equation becomes

$$0 = (0 - \tfrac{1}{2}mv^2) + (\tfrac{1}{2}kx^2 - 0)$$

The kinetic energy goes from zero to something; while we don't know the value of the arrow's final kinetic energy, we know the equation for that kinetic energy is $\frac{1}{2}mv^2$. The potential energy goes from $\frac{1}{2}kx^2$ to zero, because for a spring $x = 0$ is by definition the equilibrium position.

Whenever all the forces acting are conservative, mechanical energy is conserved. This means that energy can be changed from potential to kinetic or back, but the total mechanical energy must remain the same always. Here the initial potential energy was 30 J—that's the same $\frac{1}{2}kx^2$ calculation that we did above. These 30 J of potential energy are entirely converted to kinetic energy. So set $30\ J = \frac{1}{2}mv^2$ and solve for v. (Use the 0.10-kg mass of the arrow for m.) The speed is 24 m/s.

When you are asked to do a calculation, it's worth asking: Is the answer reasonable? One m/s is about 2 miles per hour.[6] This arrow was traveling in the neighborhood of 50 mph—about the speed of a car, but less than the speed of a professional baseball pitch.[7]

Exam Tip from an AP Physics Veteran
If the spring were instead hanging vertically instead of vibrating horizontally, this problem would be *solved the same way*. When you have a vertical spring, define the equilibrium position $x = 0$ as the place where the mass would hang at permanent rest. Then, $F = kx$ gives the *net* force on the hanging object, rather than just the force applied by the spring; and you can use $PE = \frac{1}{2}kx^2$ (not mgh) to calculate the object's potential energy.

Power

Whether you walk up a mountain or whether a car drives you up the mountain, the same amount of work has to be done on you. (You weigh a certain number of newtons, and you have to be lifted up the same distance either way!) But clearly there's something different

---

[5]The force of the bowstring is a conservative force, because the potential energy it stores is part of the mechanical energy of the bowstring-arrow system.

[6]For those of you who didn't grow up in America, 1 m/s is a bit less than 4 km/hr.

[7]For those of you who are uncomfortable with the World's Greatest Sport, this is less than the speed of a tennis player's serve.

about walking up over the course of several hours and driving up over several minutes. That difference is power.

> Power: energy/time

Power is, thus, measured in units of joules/second, also known as watts. A car engine puts out hundreds of horsepower, equivalent to maybe 100 kilowatts; whereas, you'd work hard just to put out a power of a few hundred watts.

〉 Practice Problems

Note: An additional drill involving graphical analysis of a mass on a spring is available in Chapter 18.

1. A 0.5-kg cart released from rest at the top of a smooth incline has gravitational energy of 6 J relative to the base of the incline.

    (a) Calculate the cart's speed at the bottom of the incline.
    (b) When the cart has rolled halfway down the incline, the cart's gravitational potential energy will be:
       (A) Greater than 3 J
       (B) Less than 3 J
       (C) Equal to 3 J
       Justify your answer.
    (c) When the cart has rolled halfway down the incline, the cart's kinetic energy will be
       (A) Greater than 3 J
       (B) Less than 3 J
       (C) Equal to 3 J
       (D) Unknown without knowledge of the cart's speed
       Justify your answer.
    (d) When the cart has rolled halfway down the incline, the cart's speed will be:
       (A) Half of its speed at the bottom
       (B) Greater than half of its speed at the bottom
       (C) Less than half of its speed at the bottom
    (e) A 1.0-kg cart is released from rest at the top of the same incline. At the bottom, it will be moving
       (A) Faster than the 0.5-kg cart
       (B) Slower than the 0.5-kg cart
       (C) The same speed as the 0.5-kg cart

2. A 0.1 kg stone sits at rest on top of a compressed vertical spring. The potential energy stored in the spring-earth-stone system is 40 J. The spring is released, throwing the stone straight up into the air.

    (a) How much kinetic energy will the stone have when it first leaves the spring?
    (b) How much gravitational energy, relative to the spot where the stone was released, will the stone have when it reaches the peak of its flight?
    (c) Calculate the height above the release point to which the stone travels.
    (d) Suggest something we could change about this situation that would cause the stone to reach a height double that calculated in Part (c).

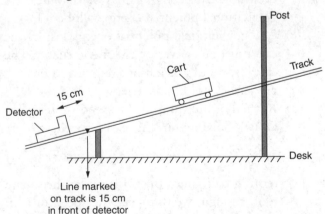

3. In the laboratory, a motion detector records the speed of a cart as a function of time, stopping its reading when the cart is 15 cm in front of the detector at the line marked on the track. The cart is released from rest at the position shown.

    (a) The kinetic energy of the cart at the line marked on the track is equal to the gravitational energy *mgh* of the cart at its initial position.

On the preceding diagram, draw and label the distance you would measure for the height *h* of the cart.

(b) Explain in some detail how commonly available laboratory equipment could be used to measure the labeled height *h*.

(c) If the height *h* were doubled in a second trial, the motion detector would read
   (A) The same speed as in the first trial
   (B) Two times the speed in the first trial
   (C) Four times the speed in the first trial
   (D) $\sqrt{2}$ times the speed in the first trial
   Justify your answer.

4. Student A lifts a 50-N box from the floor straight up to a height of 40 cm in 2 s. Student B lifts a 40-N box straight up from the floor to a height of 50 cm in 1 s.

(a) Compared to Student A, Student B does
   (A) The same work but develops more power
   (B) The same work but develops less power
   (C) More work but develops less power
   (D) Less work but develops more power
   Justify your answer

(b) Now Student A instead lifts the 50-N box from the floor diagonally, moving the box 40 cm to the right and 40 cm upward in the same 2 s.
   (A) Compared to the work he did originally, does Student A do more, less, or the same work?
   (B) Compared to the power he developed originally, does Student A develop more, less, or the same power?

> Solutions to Practice Problems

1. (a) Here gravitational energy is converted to kinetic energy. ("Smooth" generally means friction is negligible.) Kinetic energy at the bottom will be 6 J, which is equal to $\frac{1}{2}mv^2$. Plug in the 0.5-kg mass of the cart and solve for *v* to get 4.9 m/s.

   (b) Gravitational potential energy is *mgh*. Since the *h* term is in the numerator and not squared or square rooted, cutting *h* in half cuts the whole equation in half as well; so the cart's gravitational potential energy will be 3 J.

   (c) Any gravitational potential energy lost by the cart must be converted to kinetic energy. The cart lost 3 J of gravitational energy, so the cart now has 3 J of kinetic energy.

   (d) At the bottom, to find the speed we set 6 J = $\frac{1}{2}mv^2$. Solving for the cart's speed, we get $v = \sqrt{\dfrac{2 \cdot (6\text{ J})}{m}}$. We're going to cut that 6 J term in half. Since the 6 J is under the square root, though, we don't cut the speed in half; instead, we multiply the speed by $\dfrac{1}{\sqrt{2}}$. If you don't see why that gives a speed greater than half the speed at the bottom, try carrying out the entire calculation—you should get 3.5 m/s.

   (e) The energy conversion is the same—we're setting gravitational energy (*mgh*) at the top equal to kinetic energy ($\frac{1}{2}mv^2$) at the bottom. Notice the mass on both sides—the mass cancels, and so doesn't affect the result. Thus, the speed is the same for either cart. Again, feel free to do the calculation with $m = 1.0$ kg to get 4.9 m/s again.

2. (a) No forces external to the spring-earth-stone system act on the block. Therefore, mechanical energy is conserved. The 40 J of potential energy is converted to the stone's kinetic energy, or 40 J.

   (b) Again, mechanical energy is conserved. The 40 J of kinetic energy are now converted entirely to 40 J of gravitational energy.

   (c) That 40 J of gravitational energy at the peak can be set equal to *mgh*. Solve for *h* to get 40 m.

   (d) The energy conversion here is spring energy → gravitational energy. Mathematically, that's $\frac{1}{2}kx^2 = mgh$. Solving for *h*, $h = \frac{kx^2}{2mg}$. To double the height, we could use a spring with double the spring constant of the original spring, because *k* is unexponented[8] in the numerator. We could compress the spring 1.4 times its original compression, since when the *x* in the numerator is squared that would multiply the whole expression by 2. We could use a

---
[8]If "unexponented" is a word…

rock of mass 0.05 kg; with m in the denominator, halving the mass doubles the entire expression. (Okay, I suppose we could go to some new planet where g is 5 N/kg. If you will fund that trip, I'll give you credit for that answer.)

3. (a) In the equation mgh, h represents the vertical distance above the lowest position or some reference point. Here the reference point is the line on the track. The motion detector reads the front of the cart, so h must be measured to the front of the cart, not the middle or back. See above for the answer.

   (b) Use a meterstick, obviously, but it's not an easy measurement to make. First, measure the vertical distance from the desk to the line on the track 15 cm in front of the detector. Then measure the vertical distance from the desk to the track directly under the front of the cart; then subtract the two distance measurements. You can get more accurate measurements if you use a bubble level and plumb bob to ensure the table is horizontal and the measurements are vertical. (If you want to measure along the track the distance from the front of the cart to the line, use an angle measurer to get the angle of the track, then use trigonometry.)

(c) The energy conversion here is gravitational energy → kinetic energy. In equations, that's $mgh = \frac{1}{2}mv^2$. Solve for v to get $v = \sqrt{2gh}$. The variable h is in the numerator but under the square root, so doubling h multiplies the speed by the square root of 2, choice D.

4. (a) The work done by the student is equal to the change in the box's gravitational potential energy—that's mgh. The time it takes the student to lift the box doesn't depend on time at all. Plugging in, we find that Student A does (5 kg)(10 N/kg)(0.40 m) = 20 J of work on the box. Student B does (4 kg)(10 N/kg)(0.50 m) = 20 J of work, also. Now, power is work divided by the time it takes to do that work. Since they do the same amount of work, whoever takes less time to do the work develops more power. That's Student B. So the answer is choice A.

   (b) (A) As above, the work done by Student A on the box is mgh. Here h represents the vertical height above the lowest position. Since that vertical height is still 40 cm, Student A has done the same work. (You could also recognize that the horizontal displacement is not parallel to the box's weight, or to the force Student A applied to lift the box.[9])

   (B) Since the work done by Student A is the same as before, and it took the same amount of time, the power (= work/time) is the same.

⟩ Rapid Review

- Work is done when a force is exerted on an object, and that object moves parallel to the direction of the force.

- A "conservative" force converts potential energy to other forms of mechanical energy when it does work. Thus, a conservative force does not change the mechanical energy of a system.

- The only types of potential energy used in AP Physics 1 are due to gravity (mgh) and due to a spring ($\frac{1}{2}kx^2$).

- The work-energy theorem can be written as $W_{NC} = (\Delta KE) + (\Delta PE)$, where W_{NC} is the work done by a nonconservative force.

- Whenever the force on an object is not steady, energy conservation methods must be used to solve the problem. The most common of these situations are curved tracks, springs, and pendulums.

---

[9]The force Student A applies on the box is straight up, at least while the box is moving at constant speed. Once the box starts moving horizontally, no force in that direction is necessary to continue its motion—that's Newton's first law.

CHAPTER 14

Rotation

IN THIS CHAPTER

Summary: When an object rotates, the rotation obeys rules similar to those of a moving object. Most equations and concepts have analogues for rotation. Whereas mass describes an object's resistance to a change in speed, *rotational inertia* describes an object's resistance to a change in rotational speed. Rotational inertia depends on mass as well as on the distribution of that mass.

Definitions

- ✪ **Centripetal acceleration** is the name given to an object's acceleration toward the center of a circle. "Centripetal" simply means "toward the center."
- ✪ **Torque** occurs when a force applied to an object could cause the object to rotate.
- ✪ The **lever arm** for a force is the closest distance from the fulcrum, pivot, or axis of rotation to the line on which that force acts.
- ✪ Everything covered in the previous review chapters is sufficient to describe "translational" motion. When an object rotates around a central point, or when an object is itself rotating as it moves, then we need some additional concepts. Just know that each of these rotational quantities is not truly new. Each rotational quantity should be treated exactly the same way as its translational analogue: for example, if you know how to deal with linear momentum, then angular momentum applies the same ideas to rotating objects.

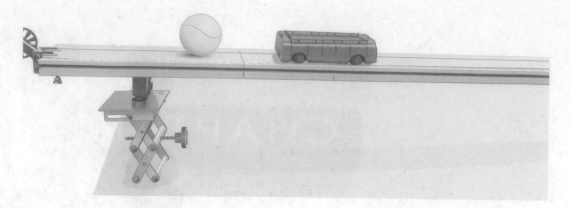

The ball rolling down the ramp in the preceding figure has translational kinetic energy because it is moving. It also has rotational kinetic energy because it is spinning. The ball is losing gravitational energy because it is changing its vertical height; that loss of potential energy is converted into a gain in kinetic energy because no nonconservative forces act on the ball.

Circular Motion

Example 1: A car of mass 1,000 kg travels at constant speed around a flat curve that has a radius of curvature of 100 m. The car is going as fast as it can go without skidding.

Does this car have an acceleration? Why yes, it does, even though it moves at constant speed. Its acceleration along its direction of motion is zero, because the car isn't speeding up or slowing down. However, its direction of motion is always changing; acceleration is technically the change in an object's vector *velocity* each second, and a change in the direction of motion is a change in velocity.

FACT: When an object moves in a circle, it has an acceleration directed toward the center of the circle. The amount of that acceleration is

$$\frac{v^2}{r}$$

Strategy: When an object is moving in a circle, often a standard Newton's second law approach is correct. Draw a free-body diagram, and then write $F_{net} = ma$ in each direction.

Draw a free-body diagram of this car as it moves. It's most useful to view this car from behind; let's say the car is turning to the right, so that the center of the circular motion is in the place indicated.

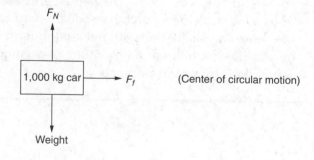

The forces are as follows:

F_N is the normal force of the road on the car.
Weight is the force of the Earth on the car.
F_f is the frictional force of the road on the car.

Wait, why is friction acting to the right? Shouldn't friction act opposite the direction of motion?

Ah. There certainly could be friction or air drag acting backwards, opposite the direction of motion; but with the car moving at constant speed, that would have to be canceled by a forward engine force.[1] Since we know the car moves at constant speed, that's probably not particularly relevant to the problem.

When a car goes around a flat curve, some sort of force must act toward the center of the circle—otherwise, the centripetal acceleration couldn't exist. How do we know it's friction in this case? Imagine the car were moving forward on a slick, flat sheet of ice. The car couldn't go around a curve at all, then; turning the wheels would do nothing. On an asphalt road, it's the static frictional force of the asphalt on the tires that pushes the car toward the center of the circle.

No matter what kind of question you're asked about this situation, the next step is to use Newton's second law in both the vertical and horizontal directions. Vertically, the car's acceleration is zero; the car isn't burrowing into the road or lifting off the road. Horizontally, we don't have a numerical value for the acceleration, but we know its equation: $\frac{v^2}{r}$

$$F_N - weight = 0$$

$$F_f = m\frac{v^2}{r}$$

The first equation tells us that the normal force on the car is equal to the car's weight of 10,000 N. In order to calculate the friction force on the car, we'd need to know one of two pieces of additional information. On one hand, if we know the car's speed, we can use the second equation to calculate F_f. On the other hand, since we know the normal force, if we knew the coefficient of static friction between the car's tires and the road, we could calculate the friction force using $F_f = \mu F_n$.

Strategy: You cannot allow yourself to become angry or frustrated when you don't have enough information to complete a calculation. Sometimes, an AP Physics 1 problem will be deliberately concocted to ask, "What additional information would you need to solve this problem?" (Remember, actual calculation on the exam will be rare.) Or, often, some seemingly necessary information will be omitted, because it will turn out that the omitted information is irrelevant.

For example, an excellent problem using this situation might *not* give the mass of the car but instead give the car's speed and ask what minimum coefficient of friction would be necessary for the car to round the curve. Pretend, say, that the car's speed is 20 m/s.

---

[1]Technically, this would be a force of static friction between the tires and the road, but that's a different day's lesson.

I can't do that. I need the mass to calculate the force of friction.

Well, true, if you needed a value for the force of friction, but that's not the question. We want the coefficient of friction.

Yeah, I know. The equation is $F_f = \mu F_n$. I need the mass to calculate the friction force, *and* since the normal force is equal to the weight, I need the mass to calculate that, too. This isn't possible.

When you're stuck with a calculation that you think needs a value that wasn't given, try just making up that value. It's likely that the unknown value will cancel out. Or, if you want to be more elegant, assign a variable to the unknown value.

Let the mass be *m*. The equations we wrote from the free-body diagram show that the friction force is $m\frac{v^2}{r}$. The normal force is the weight of the car, or the car's mass times the gravitational field *g* of 10 N/kg. Now use the equation for friction force:

$$F_f = \mu F_n$$
$$m\frac{v^2}{r} = \mu\,(mg)$$

Look at that: solve for μ, and the masses cancel. You can plug in the 20-m/s speed and the 100-m radius to get $\mu = 0.4$. A car of *any* mass can go around this curve, as long as the coefficient of friction is at least 0.4.[2]

Torque

FACT: The torque τ provided by a force is given by the equation

$$\boxed{\tau = Fd_\perp}$$

I write the symbol "⊥" by the *d* to emphasize that the distance we want is the perpendicular distance from the line of the force to the fulcrum.[3] Usually that's an easy distance to visualize.

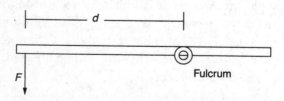

---

[2]That makes perfect sense—ever see those yellow signs warning of the appropriate speed for going around a curve? They just say, "Curve: 40 mph." They certainly *don't* say something silly like "Curve: go 10 mph for every 500 kg in your vehicle."

[3]The "fulcrum" is the point about which an object rotates, or could rotate.

What if a force isn't acting perpendicular to an extended object, like the force F on the pivoted bar that follows?

Example 2:

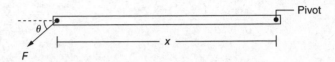

The easiest way to find the torque applied by this force is to break the force F into vertical and horizontal components.

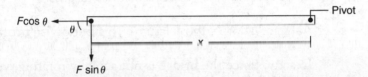

The vertical component of F applies a torque of $(F\sin \theta)x$. The horizontal component of F does not apply any torque, because it could not cause the bar to rotate. The total torque provided by the force F is just $(F\sin \theta)x$.

Lever Arm

This distance $d_\perp$ is sometimes referred to as the "lever arm" for a force. By definition, the lever arm for a force is the closest distance from the fulcrum to the line on which that force acts.

An alternate method of determining the torque applied by the force in Example 2 would be to find the lever arm instead of breaking F into components. Extend the line of the force in the diagram—now it's easy to label the lever arm as the closest distance from the pivot to the line of the force. By trigonometry, you can figure out that the lever arm distance is equal to $x\sin \theta$. No matter how you look at it, then, the torque provided by F is still $(F\sin \theta)x$.

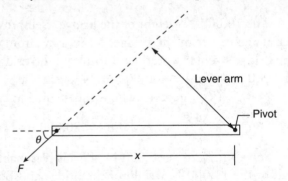

Calculations with Torque

You may be asked to calculate a force or a torque when an extended object experiences multiple forces, but generally only when that object is in equilibrium—that is, when up forces equal down forces, left forces equal right forces, and counterclockwise torques equal clockwise torques.

Example 3: Bob is standing on a bridge. The bridge itself weighs 10,000 N. The span between pillars A and B is 80 m. Bob, whose mass is 100 kg, stands 20 m from the center of the bridge as shown.

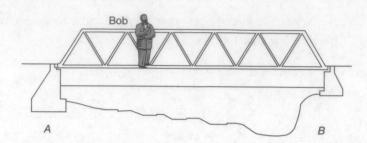

Generally, a problem with a bridge, plank, or some sort of extended object will ask you to describe or solve for the forces supporting the bridge. The approach is to make a list of torques acting in each direction, clockwise and counterclockwise, and then set the counterclockwise torques equal to the clockwise torques.

Aarrgh. Where's the fulcrum? This bridge isn't rotating anywhere!

Exactly. Since the bridge is not actually rotating, you can choose anywhere you like as the fulcrum. It's easiest in this case, to choose one of the supports as the fulcrum, because then that support provides zero torque, and the lever arm for that force would be zero.

Let's choose support A as the fulcrum. What torques do we see?

The force of Support B (I'll call it F_B) provides a torque equal to F_B(80 m), because support B is 80 m from support A. This torque is *counterclockwise*, because pushing up on the bridge pivoted at A would rotate the bridge this way ↺.

The weight of Bob provides a clockwise torque of (1,000 N)(20 m) = 20,000 m·N. (We don't use 100 kg, because that's a mass, not a force; the force acting on the bridge is due to Bob's weight.)

> **Exam Tip from an AP Physics Veteran**
> In a torque problem with a heavy extended object, just pretend that the object's weight is all hanging at the object's center of mass.

The 10,000-N weight of the bridge itself provides a torque. Pretend that all 10,000 N act at the center of the bridge, 40 m away from each support. A weight pulling down at the bridge's center would tend to rotate the bridge clockwise. The torque we want here is (10,000 N)(40 m) = 400,000 m·N, clockwise.

Now, set counterclockwise torques equal to clockwise torques:

$$F_B(80 \text{ m}) = 20,000 \text{ m·N} + 400,000 \text{ m·N}$$

Solve for F_B to get 5,300 N. This is reasonable because pillar B is supporting *less* than half of the 11,000-N weight of the bridge and Bob. Because Bob is closer to pillar A, and otherwise the bridge is symmetric, A should bear the majority of the weight—and it does.

Rotational Kinematics

An object's "rotational speed" says how fast the object rotates—that is, how many degrees or radians it rotates through per second. Rotational speed is generally given by the lowercase Greek variable omega, ω. An object's "rotational acceleration" α describes how much the rotational speed changes in one second. The variable θ represents the total angle through which an object rotates in some time period.

Just like position x, speed v, and acceleration a are related through the kinematics formulas given in Chapter 10, rotational angle, speed, and acceleration are related by the same formulas:

$$
\begin{aligned}
&1. \quad \omega_f = \omega_0 + \alpha t \\
&2. \quad \Delta\theta = \omega_0 t + \tfrac{1}{2}\alpha t^2 \\
&3. \quad \omega_f^2 = \omega_0^2 + 2\alpha\Delta\theta
\end{aligned}
$$

It is highly unlikely that you'll be asked to actually make much of a calculation with these equations. Rather, you might be asked to rank rotating objects by their angular speed or acceleration; or you might be asked, "Is it possible to calculate…." The general approach to a calculation should be identical to that for nonrotational kinematics: Make a chart with the five variables in it. If you can identify a value for three of the five variables, the problem is solvable.

Rotational Inertia

Just like "inertia" refers to an object's ability to resist changes in its motion, "rotational inertia"[4] refers to an object's ability to resist changes in its rotational motion. Two things affect an object's ability to resist rotational motion changes: the object's mass and how far away that mass is from the center of rotation.

There are three ways to figure out something's rotational inertia:

1. For a single "point" particle that is moving in a circle around an axis, its rotational inertia is given by the equation $I = mr^2$. Here, m is the mass of the particle, and r is the radius of the circle.
2. For an object with some kind of structure, like a spinning ball, a disk, or a rod, a formula for its rotational inertia will generally be given if you need it.
3. For a system consisting of several objects, you can add together the rotational inertia of each of the objects to find the total rotational inertia of a system.

Example 4: Three meter-long, uniform 200-g bars each have a small 200-g mass attached to them in the positions shown in the diagram. A person grips the bars in the locations shown and attempts to rotate the bars in the directions shown.

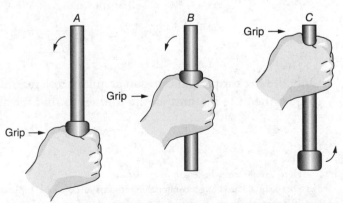

---

[4] The AP Physics 1, Algebra-Based Exam will always use the term "rotational inertia," usually represented by the variable I. Many textbooks and teachers will use the older term "moment of inertia" to refer to the same quantity. Don't be confused.

It's possible that you might be asked to calculate the rotational inertia of one of these gripped rods. If so, in the problem statement you'd be given the formula to calculate the rotational inertia of a rod: $\frac{1}{12}ML^2$ when pivoted in the center, and $\frac{1}{3}ML^2$ when pivoted at the end.[5] You know that the rotational inertia of the small mass is mr^2, where r is the distance from the mass to the grip. In each situation, add the rotational inertia of the rod to the rotational inertia of the small mass.

For Grip A, the rod's rotational inertia is $\frac{1}{3}ML^2 = \frac{1}{3}(0.2 \text{ kg})(1 \text{ m})^2 = 0.067 \text{ kg} \cdot \text{m}^2$. The small mass contributes nothing to the rotational inertia, because it is not rotating—the r term in $I = mr^2$ is zero. Thus, the total rotational inertia is 0.067 kg·m².

For Grip B, the rod is pivoted in the center, so its rotational inertia is $\frac{1}{12}ML^2 = 0.017 \text{ kg} \cdot \text{m}^2$. The small mass isn't rotating. So the total rotational inertia is just 0.017 kg·m².

Finally, for Grip C, the rod's rotational inertia is $\frac{1}{3}ML^2 = \frac{1}{3}(0.2 \text{ kg})(1 \text{ m})^2 = 0.067 \text{ kg} \cdot \text{m}^2$. The small mass's rotational inertia is $(0.2 \text{ kg})(1 \text{ m})^2 = 0.20 \text{ kg} \cdot \text{m}^2$. Thus, the total rotational inertia is the sum of both contributions, 0.27 kg·m².

Calculations aren't usually the point, though. This situation is just begging to become a ranking task: without any specific values for the masses or lengths of the items, rank the grips by their rotational inertia. As long as we know that the small mass is equal to the mass of the rod, and that the rods are equal in length, then the ranking can be done. You can see that the small mass contributes nothing to the rotational inertia in A and B without calculation.

You can see that Grip A provides a greater rotational inertia than Grip B, *even without knowing the formulas $\frac{1}{12}ML^2$ and $\frac{1}{3}ML^2$*. Reason from the properties of an object that contribute to its rotational inertia: mass, and how far away that mass is from the axis of rotation. The rods have the same mass in A and B. But Rod A has much more mass that is far away from the grip; Rod B has more mass closer to the grip. Therefore, Rod B will be easier to rotate, and Rod A will have more rotational inertia.

Then, of course, Grip C combines the "worst" of both worlds: just the rod by itself provides the same rotational inertia as in Grip A, but the mass is also contributing to the rotational inertia. The final ranking would be $I_C > I_A > I_B$.[6]

Newton's Second Law for Rotation

Just as linear acceleration is caused by a net force, *angular* acceleration is caused by a net *torque:*

$$\boxed{\tau_{net} = I\alpha}$$

Only the net torque can cause an angular acceleration. If more than one force is applying a torque, then use the sum[7] of the torques to find the angular acceleration.

---

[5]Here, L represents the length of the bar.

[6]Okay, you *do* need to get comfortable with this sort of verbal explanation of concepts that refers to equations and facts but doesn't make direct calculation. If you are confused on this sort of problem, it is okay to make up values for whatever you need, and calculate. I don't at all recommend memorizing all the different formulas for rotational inertia of a rod, sphere, hoop, disk, etc. But if you happen to remember them, it's fine to use them.

[7]Or, if the torques are acting in opposite directions, use the difference.

Example 5: A turntable of known mass and radius is attached to a motor that provides a known torque. Using the torque of the motor and the rotational inertia of the turntable in Newton's second law for rotation, then, using rotational kinematics, a student predicts that it should take 5.0 s for the turntable to speed up from rest to its maximum rotational speed. When the student measures the necessary time, though, he discovers that it takes 6.8 s to reach maximum rotational speed.

First, you should be able to describe how to perform such a measurement in your laboratory. There's a bazillion ways of doing so: the idea is to make many measurements of rotational speed until that rotational speed doesn't change. Rotational speed could be measured with a video camera and a protractor, by running the video frame-by-frame to see how many degrees the turntable advances per frame. Or tape a tiny piece of paper to the edge of the turntable, and have that paper trigger a few photogates; you can measure the angle between the photogates, and the photogates will tell you how much time it took for the turntable to traverse that angle.

This problem is setting up for you to figure out why the prediction didn't match the measurement.[8] The most obvious issue is that the torque provided by the motor might not be the *net* torque on the turntable. Friction in the bearings of the turntable could easily provide a torque in the opposite direction to that provided by the motor. Thus, the real value for net torque will be *lower* than the value the student used. And, by $\tau_{net} = I\alpha$, the real angular acceleration will be lower; finally, by $\omega_f = \omega_0 + \alpha t$, a smaller angular acceleration to get to the same final speed means that the time t will be longer than predicted.

Angular Momentum

The rotational analogue of the impulse-momentum theorem involves torque and angular momentum rather than force and linear momentum:

$$\Delta L = \tau \cdot \Delta t$$

Here τ is the net torque acting on an object, and Δt is the time during which that torque acts. The change in the object's angular momentum is ΔL.

An object's angular momentum can be calculated using three methods:

1. For a single "point" particle that is moving in a circle around an axis, its angular momentum is given by $L = mvr$. Here, r represents the radius of that circle.
2. For a single "point" particle that is moving in a straight line,[9] its angular momentum is also given by $L = mvr$; but in this case r represents the "distance of closest approach"

---

[8]Please don't automatically say "human error." There's no such thing as "human error," and using that phrase is basically an automatic *wrong* on the AP Exam.

[9]Angular momentum must always be defined with respect to some central axis of rotation. For most rotating objects, that axis is obvious. For a particle moving in a straight line, you have to say what position you're calculating angular momentum for, but the particle can still have angular momentum.

from the line of the particle's motion to the position about which angular momentum is calculated, as shown below.

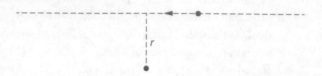

3. For an extended object with known rotational inertia I, angular momentum is given by $L = I\omega$.

For most problems, you can reason by analogy to the linear impulse-momentum theorem. Just as a force applied for some time will change an object's linear momentum, a torque applied for some time will change an object's angular momentum. Just as the area under a force versus time graph gives the change in an object's linear momentum, the area under a torque versus time graph gives the change in an object's angular momentum.

Conservation of Angular Momentum

FACT: In any system in which the only torques acting are between objects in that system, angular momentum is conserved. This effectively means that angular momentum is conserved in *all* collisions, but also in numerous other situations.

Example 6: A uniform rod is at rest on a frictionless table. A ball of putty, whose mass is half that of the rod, is moving to the left, as shown. The ball of putty collides with and sticks to the rod.

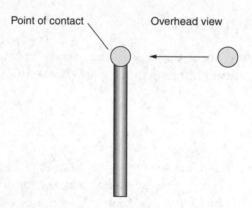

Exam Tip from an AP Physics Veteran
Usually, the fulcrum or axis of rotation is obvious. But when an object is not forced to pivot at some specific position, if it rotates it will most likely rotate about its center of mass.

Start with what quantities are conserved for the putty-rod system. It's a collision, in which the only forces involved are the force of the putty-on-rod and rod-on-putty. Therefore, linear momentum is conserved. Similarly, the putty applies a torque to the rod because it pushes on the rod at a position away from its center of mass. But the only torques involved are provided by objects in the system, so angular momentum is also conserved.

This collision cannot be elastic because the putty sticks to the rod; so kinetic energy was not conserved. Mechanical energy was also not conserved because the kinetic energy of the putty was not stored as potential energy in a spring or gravitational field. Of course, the sum of *all* forms of energy was conserved, because whatever kinetic energy was lost by the putty-rod system was converted to microscopic internal energy, and thus the temperature of the putty-rod will increase.

Where's the center of mass? Using the equation from Chapter 12, call $x = 0$ the top of the rod. Pretend the rod is 1-m long and 1 kg in mass. Then the putty is 0.5 kg in mass. So $m_{putty}(0) + m_{rod}(0.5 \text{ m}) = m_{total}(x_{cm})$.[10] Plugging in the masses, you get $x = 0.33$ m.

But the problem emphatically did *not* say that the rod was 1-m long.

Right. Whatever the rod's length, its center of mass is one-third of the way down the rod.

Exam Tip from an AP Physics Veteran
When you're asked about the center of mass speed, you can ignore all angular stuff. In that case, just treat the collision as if these were carts colliding.

Again, you can make up values to find the speed of the center of mass after collision. If the putty's initial speed were 1 m/s, then the total momentum before collision is 0.5 N·s. By conservation of momentum, that's also the total momentum after collision, but the mass of the combined objects after collision is 1.5 kg. The speed of the center of mass would be 0.33 m/s. But since the initial speed wasn't given, you can only definitively say that the speed of the center of mass after collision will be one-third of the speed of the putty before collision.

Conservation of Angular Momentum Without Collisions

FACT: Angular momentum is conserved any time an object, or system of objects, experiences no net torque.

Example 7: A person stands on a frictionless turntable. She and the turntable are spinning at one revolution every two seconds.

Sure, the person can wiggle and exert a torque on the turntable. But this torque is internal to the person-turntable system. The person thus cannot change the angular momentum of the person-turntable system.

This doesn't mean that she can't change her angular speed. What if she throws her arms way out away from her body?[11] Her rotational inertia would change, because she'd have the same total mass but more of that mass would be far away from the center of rotation. Her angular momentum can't change. By $L = I\omega$, to keep a constant L with a bigger I, the angular speed ω must decrease. This is the physical basis for how figure skaters can control their spinning.

---

[10]It's (0.5 m) because the center of mass of the rod by itself must be halfway down the rod.

[11]AP reader Matt Sckalor quite reasonably asks, "After she throws the first arm, what part of her body does she use to throw the other arm?" Perhaps I should say she "extends" her arms.

Example 8: A planet orbits a sun in an elliptical orbit.

You won't have to deal with elliptical orbits in the sense of making calculations, or using Kepler's Laws and Physics C-style calculations. But you can understand that angular momentum of a planet orbiting a sun must be conserved. Since the force of the sun on a planet is always on a line toward the sun itself, this force cannot provide any torque—there's no lever arm. With no torque exerted on a planet, that planet cannot change its angular momentum. Treat the planet as a point particle; its angular momentum is $L = mvr$, where r is the distance from the sun. Whenever r is big, v must be small to keep L constant. The farther away the planet is from the sun, the slower it moves.

Rotational Kinetic Energy

FACT: When an object is rotating, its rotational kinetic energy is

$$\tfrac{1}{2}I\omega^2$$

In Example 6, the putty-rod system has *both* rotational *and* linear kinetic energy after the collision. The total kinetic energy is then the sum of $\tfrac{1}{2}mv^2$, where v is the speed of the center of mass, plus the rotational kinetic energy.

When an object rotates, the work-energy theorem still applies. The kinetic energy terms each include the addition of rotational and linear kinetic energy. In Example 6, work is done by a nonconservative force (the force of the rod on the putty). To find out how much work was done, set $W_{NC} = \Delta KE + \Delta PE$. This situation doesn't involve a spring or a changing vertical height, so $\Delta PE = 0$. The kinetic energy was originally just $\tfrac{1}{2}mv^2$ for the putty; after the collision, the kinetic energy is as discussed in the previous paragraph, the speed of the center of mass plus the rotational kinetic energy.

› Practice Problems

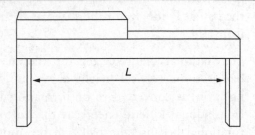

1. A uniform wooden block has a mass m. On it is resting half of an identical block, as shown above. The blocks are supported by two table legs, as shown.

   (a) Which table leg, if either, should provide a larger force on the bottom block? Answer with specific reference to the torque equation.

   (b) In terms of given variables and fundamental constants, what is the force of the right-hand table leg on the bottom mass?

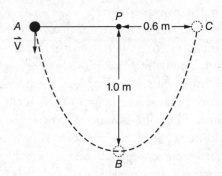

2. A small ball of mass m moving on a frictionless horizontal surface is attached to a rubber band whose other end is fixed at point P. The ball moves along the dotted line in the preceding figure, stretching the rubber band. When it passes Point A, its velocity is v directed as shown.

   (a) Is the angular momentum of the ball about Point P conserved between positions A and B?

   (b) Is the linear momentum of the ball conserved between positions A and B?

   (c) Describe a system in this problem for which mechanical energy is conserved as the ball moves from A to B.

   (d) Explain why the net force on the ball at Point B is not $\frac{mv^2}{(1.0 \text{ m})}$.

3. A smooth, solid ball is released from rest from the top of an incline, whose surface is very rough. The ball rolls down the incline without slipping.

   (a) Describe in words the energy conversion for the ball from its release until it reaches the bottom of the incline.

   (b) Is the mechanical energy of the ball-Earth system conserved during its roll?

   (c) This ball is replaced by a new ball, whose surface and mass are identical to the first ball, but which is predominantly hollow inside. Describe any differences in its roll down the incline without slipping, with explicit reference to forms of energy.

› Solutions to Practice Problems

1. (a) Call the force of the left support F_L, and the force of the right support F_R. Consider the middle of the bottom block as the fulcrum. Then one clockwise torque acts: $F_L \cdot (L/2)$. Two counterclockwise torques act, though: $F_R \cdot (L/2)$ and $(1/2)m \cdot (1/4)L$. The point is that if you have to add something to the torque provided by the right support to get the torque provided by the left support, the left support thus provides more torque. Because the supports are the same distance from the center, the left support provides more force, too.

   (b) You certainly could use the reasoning in Part (a) with the fulcrum in the center, along with the total support force equaling $1.5Mg$ (vertical equilibrium of forces). However, it's much easier mathematically to just call the left end of the rod the fulcrum. Then the counterclockwise torque is $F_R \cdot L$. The clockwise torque is $(1/2)mg \cdot (L/4) + mg \cdot (L/2)$. Set these equal and play with the fractions to get $F_R = (\frac{1}{8} + \frac{1}{2})mg = \frac{5}{8}mg$.

2. (a) Angular momentum is conserved when no torques external to the system act. Here the system is just the ball. The only force acting on the ball is the rubber band, which is attached to Point P. The force applied by the rubber band can't have any lever arm with respect to P and thus provides no torque about point P, so the ball's angular momentum about point P can't change. Angular momentum is conserved.

   (b) Linear momentum is conserved when no forces external to the system act. Here the system is just the ball. The rubber band is external to the system and applies a force; therefore, linear momentum is *not* conserved.

   (c) Mechanical energy is conserved when no force external to the system does work. The rubber band does work on the ball, because it applies a force and stretches in a direction parallel to the force it produces. Consider the rubber band part of the system. The post at Point P still applies a force to the ball–rubber band system, but since the post doesn't move, that force does no work on the ball–rubber band system. Any kinetic energy lost by the ball will be stored as elastic energy in the rubber band. The mechanical energy of the ball–rubber band system is conserved.

   (d) The general form of this equation is fine—the ball's path at Point P is, at least in the neighborhood of P, approximately circular. The ball experiences a centripetal acceleration at Point B, and centripetal acceleration is v^2/r. The problem is that if the r term is 1.0 m, then the v term must represent the speed at Point B. With angular momentum conserved, the total of mvr must always be the same. The ball's mass doesn't change. The distance r from Point P gets bigger from A to B, so the speed must get smaller. The equation given uses the given variable v which represents the speed of the ball at Point A, not the speed at B, and so is invalid.

3. (a) Gravitational energy at the top (because the ball is some vertical height above its lowest position) is converted to both rotational and translational kinetic energy at the bottom—rotational because the ball will be spinning, and translational because the ball's center of mass will move down the incline.

   (b) Mechanical energy is conserved when no nonconservative forces act. Here the Earth's gravitational field can give the ball kinetic energy, but since the Earth is part of the system and since the gravitational force is conservative, that still allows for conservation of mechanical energy. Friction is a nonconservative force, but here friction does no work.

   (c) The hollow ball of the same mass will have greater rotational inertia, because the mass is concentrated farther from the center of rotation. The ball's gravitational energy before the rolling begins is the same as the previous scenario, because the height of the incline is the same. The total kinetic energy at the bottom will not change; the question is how much of that kinetic energy will be rotational, and how much will be translational.

   Rotational KE is $\frac{1}{2}I\omega^2$; the angular speed ω depends on the translational speed v. (The faster the ball is moving, the more it's rotating, too.) Therefore, rotational kinetic energy depends on v^2. Translational kinetic energy also depends on v^2 in the formula $\frac{1}{2}mv^2$. The hollow ball has bigger I. The speed v must be lower for the hollow ball so that $\frac{1}{2}I\omega^2 + \frac{1}{2}mv^2$ adds to the same value for both balls.

❯ Rapid Review

- When an object moves in a circle, it has an acceleration directed toward the center of the circle. The amount of that acceleration is $\frac{v^2}{r}$.

- The torque τ provided by a force is given by the equation $\tau = Fd_\perp$.

- In any system in which the only torques acting are between objects in that system, angular momentum is conserved. This effectively means that angular momentum is conserved in *all* collisions, but also in numerous other situations.

- Angular momentum is conserved any time an object, or system of objects, experiences no net torque.

- When an object is rotating, its rotational kinetic energy is $\frac{1}{2}I\omega^2$.

CHAPTER 15

Gravitation

IN THIS CHAPTER

Summary: The force on an object due to gravity is *mg*, where *m* is mass and *g* is the gravitational field. The gravitational field produced by an object of mass *m* is $G\dfrac{M}{d^2}$, where *d* is the distance from the object's center. Want to be sure you know the difference between a gravitational field, a gravitational force, and the universal gravitation constant? This chapter explains these concepts.

Definitions

- ✪ The **gravitational field** *g* near a planet tells how much 1 kg of mass weighs at a location. Near Earth's surface, the gravitational field is 10 N/kg.
- ✪ The **gravitational force** of a planet on any other object in the planet's gravitational field is *mg*, where *m* is the mass of the object experiencing the force.
- ✪ **Newton's gravitation constant** is the universal constant $G = 6 \times 10^{-11} \text{ N·m}^2/\text{kg}^2$.
- ✪ The **free-fall acceleration** (sometimes imprecisely called the **acceleration due to gravity**) near a planet is, by an amazing coincidence of the universe, equal to the gravitational field near that planet. Near Earth, then, the free-fall acceleration is 10 m/s per second because the gravitational field is 10 N/kg.

The gravitational force is the weakest of the fundamental forces in nature. However, when enormous amounts of mass congregate—as in a star or a planet—the gravitational force becomes dominant. The picture as follows gives a hint of the scales involved in studying stars and planets—it's the Earth and the Moon, but drawn to approximately the proper scale. Sure, the Earth seems humongous, especially when you're caught in traffic. But *all* planets and stars seem small when the distances between them are considered.

Earth   Moon

The word *gravity* by itself is an ambiguous term. Begin this chapter by carefully reading the differences between all the various things that could be referred to by the word *gravity*. See the four preceding definitions, each of which relate to the word.

Determining the Gravitational Field

The gravitational field is a vector quantity—this means it has an amount and a direction. The direction is always toward the center of the Earth (or whatever is creating the gravitational field).

FACT: The amount[1] of gravitational field depends on two things: the mass of the planet creating the field (M) and the distance you are from that planet's center (d). The relevant equation for the gravitation field g produced by a planet is

$$g = G\frac{M}{d^2}$$

Some books, and probably even the AP Exam, will use the variable r for the distance from the planet's center. That's fine, but know that this r does *not* necessarily stand for the radius of the planet—it means the distance from the planet's center.

Example 1: A 20-kg rover sits on newly discovered Planet Z, which has twice the mass of Earth and twice the diameter of Earth.

You do not get a table of astronomical information on the AP Physics 1, Algebra-Based Exam. Nevertheless, you might well be asked to calculate the gravitational field near the surface of Planet Z. How can that be done without knowing the mass of Planet Z? You're expected to be fluent in semiquantitative reasoning.

Even though you don't know the value of the mass of the Earth, you know that the mass of Planet Z is twice Earth's mass. Whatever the exact mass, the numerator in the gravitational field equation will double for Planet Z.

The surface of a planet is one planet-radius away from the planet's center; so here, d means the radius of the planet. Planet Z's diameter is twice Earth's, which also means Z's radius is twice Earth's. The d term in the denominator is doubled; because d is squared, the entire denominator is multiplied by 2^2, which is 4.

---

[1]The exam will refer not to the "amount," but to the "magnitude" of a vector quantity. Just translate in your head.

Combining these effects of mass and radius, the numerator is multiplied by 2, the denominator multiplied by 4, so the entire gravitational field of the Earth is multiplied by one-half. We know Earth's gravitational field—that's 10 N/kg. Planet Z produces a gravitational field of 5 N/kg at the surface. No calculator is necessary.

The formula shows that the gravitational field produced by a planet drops off rapidly as you get far from the planet's center: if you double your distance from the planet's center, then you cut in one-fourth the value of the gravitational field.

This seems easy enough, but think about reality for a moment. Just how often do you double your distance from the center of the Earth? The radius of the Earth is about 4,000 miles. Even when you fly in an airplane, you're no more than about seven miles above the surface; so your distance from the center of the Earth is *still* about 4,000 miles.

The point is, unless you're an astronaut, the gravitational field near the surface of a planet is a constant value. Don't be tricked by the d^2 in the denominator—that only matters when you're considering objects in space.

Determining Gravitational Force

FACT: The weight of an object—that is, the gravitational force of a planet on that object—is given by mg.

That 20-kg rover would weigh 200 N on Earth (20 kg times 10 N/kg). But on Planet Z, the rover weighs only 100 N (that's 20 kg times 5 N/kg).

A weight of 100 N means that Planet Z pulls the rover downward with 100 N of force. What about the force of the rover on Planet Z? That's got to be so small it's negligible, right?

Wrong. Newton's Third Law says that the force of Planet Z on the rover is equal to the force of the rover on Planet Z. The rover pulls up on Planet Z with a force of 100 N.

Now, as you might suspect, you'd never notice or measure any effect from the rover's 100-N force on Planet Z. Planet Z is enormously massive—in the neighborhood of 10^{24} kg. By $F_{net} = ma$, you can calculate that the *acceleration* provided to the planet is immeasurably small.[2] It's the force that's the same, and the acceleration that's different.

Force of Two Planets on One Another—Order of Magnitude Estimates

Example 2: The Earth has a mass of 6.0×10^{24} kg. The Sun has a mass of 2.0×10^{30} kg. The Earth orbits the sun in a circle of radius 1.5×10^{11} m.

FACT: The gravitational force of one object on another is given by

$$F = \frac{Gm_1m_2}{d^2}$$

---

[2]And the rover's force on the planet is certainly not the *net* force on the planet, so this $F_{net} = ma$ calculation is silly anyway.

So it seems straightforward, if calculator intensive, to calculate the force the Sun exerts on the Earth (or vice versa). Just plug in the numbers. But that's *not* a likely AP Physics 1 exercise! No one cares whether you can use the buttons on your calculator correctly.

Instead, you might be asked, "Which of the following is closest to the force of the Sun on the Earth?"

(A) 10^{12} N
(B) 10^{22} N
(C) 10^{32} N
(D) 10^{42} N

Look how far apart these answer choices are. Don't use a calculator—instead make an *order of magnitude estimate*. Plug in just the powers of 10, and the answer will leap off the page. Leaving out the units of each individual term for simplicity, in the equation $F = \dfrac{Gm_1 m_2}{d^2}$ we have

$$\frac{(10^{-11})(10^{24})(10^{30})}{(10^{11})^2}$$

That's easy to simplify without a calculator—add exponents in the numerator, and then subtract the exponents in the denominator.

$$\frac{(10^{43})}{(10^{22})} = 10^{21}\,\text{N}.$$

But that's not one of the choices.

The only reasonable choice, though, is (B) 10^{22} N. The others are at least a factor of a billion too big or too small. Sure, you could have spent five minutes plugging in the more precise values into your calculator,[3] getting 3.3×10^{22} N as the answer. The order of magnitude estimate is as precise as would ever be necessary on the AP Physics 1 exam, and it's a lot easier, too.

Gravitational Potential Energy

Near the surface of the Earth, the potential energy provided by the gravitational force is

$$\boxed{PE_{gravity} = mgh}$$

That's plenty good enough for calculations with everyday objects.

However, if you're talking about objects way out in space, the gravitational potential energy possessed by two objects near one another is given by

$$\boxed{PE_{gravity} = G\frac{M_1 M_2}{d}}$$

---

[3]And then spend another 10 minutes swearing at the calculator because you left out a parenthesis, or you forgot to type a negative sign, or you forgot a decimal.

This equation is usually written with a negative sign; that negative sign simply indicates that this potential energy is *less* than zero; and in outer-space situations, zero potential energy means the objects are infinitely far away from each other.

This gravitational potential energy can be converted into kinetic energy—if two planets move toward one another due to their mutual gravitational attraction, you might be able to figure out how fast they move by calculating the gravitational potential energy possessed before and after they move. Any lost potential energy was converted into kinetic energy.

Gravitational and Inertial Mass

Example 3: Neil Armstrong had a mass of 77 kg when he went to the moon. The gravitational field on the Moon is one-sixth that on Earth.

The term "mass" is often colloquially defined as the amount of "stuff" in an object. You're likely to see an AP question of the form, "What was Neil Armstrong's mass on the Moon?" The answer is, still 77 kg. Sure, Neil's weight on the Moon was smaller than his weight on Earth, because the gravitational field on the Moon is smaller, but since he didn't cut off his leg or go on a starvation diet, his mass didn't change.

> **KEY IDEA**

FACT: Gravitational mass indicates how an object responds to a gravitational field.

FACT: Inertial mass indicates how an object accelerates in response to a net force.

FACT: In every experiment ever conducted, an object's gravitational mass is equal to its inertial mass.

The AP Exam requires you to distinguish between the two meanings of mass. Simply put, if there's acceleration involved, you're talking about inertial mass. You'll be asked to design experiments to measure each type of mass.

So let's say you want to figure out who has more *gravitational* mass, you or Neil Armstrong. Just put each of you on a balance scale—the one with the bigger scale reading has more gravitational mass. You could use a spring scale, too—whoever compresses the springs more experiences more gravitational force in the same gravitational field, and so has more gravitational mass.

But to figure out experimentally who has more *inertial* mass, you'd put each of you in an identical buggy. Speed up each buggy using the same net force for the same amount of time. Whichever of you has sped up by more—that is, whichever experienced the greater acceleration under the same net force—has the smaller inertial mass.

And if your experiment gives contradictory results—say, that Neil has more inertial mass but that you have more gravitational mass—then you should reject that result as ridiculous. An exam question might pose just this type of situation in which you have to reject impossible results.[4]

Fundamental Forces: Gravity Versus Electricity

The gravitational force is one of several "fundamental" forces in nature. The only other fundamental force on the AP Exam is the electrical force. In this context, "fundamental" means that all forces on the AP Exam are manifestations of one of these two forces.

---

[4]Well, if you really have repeatable and unassailable results of this nature, you should publish. I guarantee you'll win a Nobel Prize for your work, which is a bit more likely to earn college credit than a 5 on the AP Exam.

Oh yeah? What about, say, friction?

All of the everyday forces that result from "contact" between two objects are fundamentally electrical. The electron shells in the outermost atomic layers of the objects interact, causing friction, normal forces, tensions, punches, etc. Really. But AP Physics 1 doesn't care at all about the details of these interactions; just know that they're electrical.

As discussed in Chapter 16, the electrical force between two particles is generally much, much stronger than the gravitational force between them. The AP Exam is likely to ask you to explain why gravity is such a dominant everyday force, then.

The answer is twofold. First of all, consider the overwhelmingly massive scale of the Earth. It's tens of thousands of miles in circumference, and a trillion trillion kilograms in mass. And even with that much stuff in it, the gravitational force it exerts on you is equal to the (electrical) contact force of the ground pushing up on you.

The bigger issue is that masses always cause a gravitational field that attracts other masses—there's no such thing as a repulsive gravitational force. Electrical forces, though, can be both repulsive and attractive, depending on the sign of the charges interacting. Since in general the Earth is electrically neutral, all that unimaginably immense stuff in the Earth has very little net electrical effect, leaving gravity to keep you firmly attached to the ground.

› Practice Problems

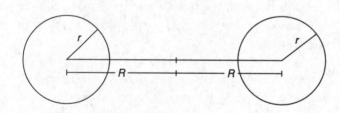

1. Two stars, each of mass M, form a binary system. The stars orbit about a point a distance R from the center of each star, as shown in the diagram above. The stars themselves each have radius r.

   (a) In terms of given variables and fundamental constants, what is the force each star exerts on the other?

   (b) In terms of given variables and fundamental constants, what is the magnitude of the gravitational field at the surface of one of the stars due only to its own mass?

   (c) In terms of given variables and fundamental constants, what is the magnitude of the gravitational field at the midpoint between the stars?

   (d) Explain why the stars don't crash into each other due to the gravitational force between them.

2. A space shuttle orbits Earth 300 km above the equator.

   (a) Explain why it would be impractical for the shuttle to orbit 10 km above the Earth's surface (about 1 km higher than the top of Mount Everest).

   (b) A "geosynchronous orbit" means that the shuttle will always remain over the same spot on Earth. Explain and describe the calculations you would perform in order to determine whether this orbit is geosynchronous. You should not actually carry out the numerical calculations, just describe them in words and show them in symbols.

   (c) The radius of Earth is 6,400 km. At the altitude of the space shuttle, what fraction of the surface gravitational field g does the shuttle experience?

   (d) When the shuttle was on Earth before launch, the shuttle's mass (not including any fuel) was 2×10^6 kg. At the orbiting altitude, what is the shuttle's mass, not including fuel?

3. A satellite is in circular orbit around an unknown planet. A second, different satellite also travels in a circular orbit around this same planet, but with an orbital radius four times larger than the first satellite.

(a) Explain what information must be known in order to calculate the speed the first satellite travels in its orbit.

(b) Compared to the first satellite, how many times faster or slower is the second satellite's speed?

(c) Bob the bad physics student says:
"The gravitational force of the planet on a satellite in circular orbit depends inversely on the orbital radius squared. Since the second satellite's orbital radius is four times that of the first satellite, the second satellite experiences one-sixteenth the gravitational force that is exerted on the first satellite."
Explain what is wrong with Bob's explanation.

4. A spacecraft is positioned between the Earth and the Moon such that the gravitational forces on the spacecraft exerted by the Earth and the Moon cancel.

(a) Is this position closer to the Moon, closer to the Earth, or halfway in between?

(b) Are the gravitational forces on the spaceship (the force exerted by the Moon, and the force exerted by the Earth) a Newton's third law force pair?

❭ Solutions to Practice Problems

1. (a) The term in the denominator of Newton's law of gravitation refers to the distance between the centers of the two stars. That distance is given as $2R$. So the answer is $G\dfrac{M^2}{(2R)^2}$ or $G\dfrac{M^2}{4R^2}$. Notice you must use the notation given in the problem—this means a capital R here. (The M is squared because the equation for gravitational force multiplies the masses of the stars applying and experiencing the force. Since the masses of both stars are the same, you get $M{\cdot}M = M^2$.)

(b) When calculating gravitational field at the surface of a star, the term in the denominator is the star's radius. That's r here. So $G\dfrac{M}{r^2}$. (The M is *not* squared here because the field is produced by a single star, not an interaction between two stars.)

(c) The left-hand star provides a field of $G\dfrac{M}{R^2}$ at the location of the midpoint, pointing toward the left-hand star. The right-hand star also provides a field of $G\dfrac{M}{R^2}$, pointing toward the right-hand star. These fields with the same amount but opposite direction cancel out when they add as vectors, producing a net gravitational field of zero at the midpoint.

(d) The direction of a force is not the same as the direction of an object's motion. Here, at any time a star is moving along an orbit, tangent to the radius of the orbit. The force applied by the other star will always be toward the center of the orbit, perpendicular to the direction of the star's motion. When a force is applied perpendicular to an object's velocity, the result is circular motion. The centripetal force changes the direction (not the amount) of the star's speed, but the force itself always changes direction so that it is pulling toward the center of the circular motion.

2. (a) The shuttle must be above the atmosphere in order to maintain a circular orbit without continually burning fuel. If the Earth had no atmosphere, then a satellite could orbit at any distance from the surface that's greater than the tallest mountain. But air resistance in the atmosphere does work on the shuttle, reducing its mechanical energy. At 300 km above the surface, though, the shuttle is above the atmosphere and experiences no forces in or against the direction of travel.

(b) This question requires a calculation: the force on the shuttle is $G\dfrac{M_{earth}M_{shuttle}}{d^2}$, where d is the distance from the center of the Earth to the location of the shuttle. Since this force is a centripetal force, we can set it equal to $\dfrac{M_{shuttle}v^2}{d}$, where v is the speed of the shuttle in orbit. If the orbit is geosynchronous, that doesn't mean that the shuttle goes the same speed as a position on Earth, it means that the *period* of the orbit is 24 hours—it goes around the Earth in the same time as a location on the surface does.

To determine the period of the orbit, set the speed equal to the orbit's circumference ($2\pi d$) divided by the period, which I'll call T. Here's how our equation looks now:

$$G\frac{M_{earth}M_{shuttle}}{d^2} = \frac{M_{shuttle}\left(\dfrac{2\pi d}{T}\right)^2}{d}.$$

There is lots of algebra here, but notice that all values are things that can be looked up: G, the mass of the Earth, the mass of the shuttle (which cancels anyway), and d, which is the radius of the Earth plus 300 km. Solve this equation for T. If the value of T turns out to be 24 hours, then the orbit is geosynchronous; if not, the orbit is not geosynchronous.[5]

(c) Your first instinct might be that you need the mass of the Earth to answer this question, because the equation for the gravitational field is $G\dfrac{M}{R^2}$. On one hand, we could calculate the mass of Earth knowing that the gravitational field near the surface is 10 N/kg—just plug in the given value of R and G from the table of information. That will work. It's more elegant to solve in variables: The fraction we want is

$$\frac{G\dfrac{M}{(6700 \text{ km})^2}}{G\dfrac{M}{(6400 \text{ km})^2}}.$$

The G and M cancel, whatever their value. Work the improper fraction to get that the gravitational field 300 km above the surface is $\dfrac{(6400 \text{ km})^2}{(6700 \text{ km})^2} = 91\%$ of g at the surface.

(d) Mass is the amount of "stuff" in an object. It doesn't matter where that object is in the universe, 1 kg of mass is 1 kg of mass. Unless

the shuttle loses a wing, its mass is still 2×10^6 kg.

3. (a) Set the gravitational force on the satellite equal to the formula for centripetal force:

$$\frac{GM_{planet}M_{satellite}}{d^2} = \frac{M_{satellite}v^2}{d}.$$

We'll want to be able to solve for v. The mass of the satellite cancels, and G is a universal constant. But we'll need to know d, the distance of the satellite's orbit from the center of the planet. And we'll need to know the mass of the planet or the period of the orbit.

(b) When we solve for v, we get $v = \sqrt{\dfrac{GM_{planet}}{d}}$. The second satellite has four times the orbital radius, which is represented by d. Multiplying by four in the denominator under the square root multiplies the whole expression by one-half. The second satellite's speed is one-half as large.

(c) The equation for the force of the planet on the satellite is $F = G\dfrac{M_{planet}M_{satellite}}{d^2}$. Sure, Bob is right about the inverse square dependence on d, but he's assumed that the satellites have the same masses as each other. If they do, Bob is correct; if not, then, the mass of the satellite shows up in the numerator of the force equation.

4. (a) The relevant equation here comes from setting the forces on the spacecraft (of mass m) equal:

$$G\frac{M_{earth}m}{d_{earth}^2} = G\frac{M_{moon}m}{d_{moon}^2}.$$

The mass of the spacecraft cancels, as does the G. Since the mass of the Earth is bigger than the mass of the Moon, the equation shows that the distance of this location from Earth d_{earth} must be larger than the distance of the location from the Moon d_{moon}. The location is closer to the Moon.

(b) A Newton's third law force pair cannot act on the same object. The force of the Moon on the spacecraft is paired with the force of the spacecraft on the Moon; the force of the Earth on the spacecraft is paired with the force of the spacecraft on the Earth.

---

[5]If you do all the plugging and chugging—which are *not* necessary—you'll find that the orbit is not geosynchronous. Geosynchronous orbits are somewhere in the 35,000-km range above Earth's surface.

〉 Rapid Review

- The amount of gravitational field depends on two things: the mass of the planet creating the field (M) and the distance you are from that planet's center (d). The relevant equation for the gravitation field g produced by a planet is $g = G\dfrac{M}{d^2}$.

- The weight of an object—that is, the gravitational force of a planet on that object—is given by mg.

- The gravitational force of one object on another is given by $F = \dfrac{Gm_1 m_2}{d^2}$.

- Gravitational mass indicates how an object responds to a gravitational field.

- Inertial mass indicates how an object accelerates in response to a net force.

- In every experiment ever conducted, an object's gravitational mass is equal to its inertial mass.

Electricity: Coulomb's Law and Circuits

IN THIS CHAPTER

Summary: Coulomb's law says that the force of one charged particle on another is $k\dfrac{q_1 q_2}{d^2}$. AP Physics 1 deals only with electrical forces between exactly two charges. But AP Physics 1 does require an understanding of direct-current circuits, including series resistors, parallel resistors, and combinations thereof. This chapter shows you how to deal with circuits, both using conceptual descriptions and using calculations.

Definitions

- ✪ Electric **charge** (Q) exists due to excess or deficient electrons on an object. Charge comes in two kinds: positive and negative. The unit of charge is the **coulomb**.
- ✪ Electric **current** (I) is the flow of (positive) charge per second. The units of current are **amperes.** One ampere means one coulomb of charge flowing per second.
- ✪ **Resistance** (R), measured in ohms (Ω), tells how difficult it is for charge to flow through a circuit element.
- ✪ **Resistivity** (ρ) is a property of a material, which implies what the resistance would be of a meter-cube bit of that material.
- ✪ **Voltage** is electrical potential energy per coulomb of charge.
- ✪ Resistors are connected in **series** if they are connected in a single path.
- ✪ Resistors are connected in **parallel** if the path for current divides, then comes immediately back together.

The AP Physics 1, Algebra-Based Exam requires you to learn about two aspects of electricity. First, you must understand how charged objects apply forces to each other in isolation, as when a balloon sticks to the wall. Next, you need to know about circuits, in which gazillions of submicroscopic flowing charges produce effects that can be measured and observed. The following picture shows such a circuit, in which flowing charge causes two bulbs to light up. The meter is positioned to read the voltage across one of the bulbs.

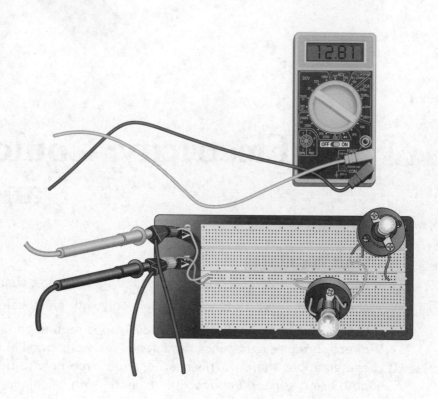

Electric Charge

All matter is made up of three types of particles: protons, neutrons, and electrons. Protons have an intrinsic property called "positive charge." Neutrons don't contain any charge, and electrons have a property called "negative charge."

The unit of charge is the coulomb, abbreviated C. One proton has a charge of 1.6×10^{-19} coulombs.

Most objects that we encounter in our daily lives are electrically neutral—things like couches, for instance, or trees, or bison. These objects contain as many positive charges as negative charges. In other words, they contain as many protons as electrons.

When an object has more protons than electrons, though, it is described as "positively charged"; and when it has more electrons than protons, it is described as "negatively charged." The reason that big objects like couches and trees and bison don't behave like charged particles is because they contain so many bazillions of protons and electrons that an extra few here or there won't really make much of a difference. So even though they might have a slight electric charge, that charge would be much too small, relatively speaking, to detect.

Tiny objects, like atoms, more commonly carry a measurable electric charge, because they have so few protons and electrons that an extra electron, for example, would make a big difference. Of course, you can have very large charged objects. When you walk across a carpeted floor in the winter, you pick up lots of extra charges and become a charged object yourself . . . until you touch a doorknob, at which point all the excess charge in your body travels through your finger and into the doorknob, causing you to feel a mild electric shock.

Electric charges follow a simple rule: *Like charges repel; opposite charges attract.* Two positively charged particles will try to get as far away from each other as possible, while a positively charged particle and a negatively charged particle will try to get as close as possible.

Only two types of charge exist. If a question on the exam purports to give evidence of a third type of charge, reject that evidence; if a question suggests a kind of charge that repels *both* positive and negative charges, reject that suggestion as silly.

Coulomb's Law

Just *how much* do charged objects attract and repel? Coulomb's law tells how much.

FACT: Coulomb's law is an equation for the force exerted by one electrical charge on another:

$$F = k \frac{Q_1 Q_2}{d^2}$$

The Qs represent the amount of charge on each object; the d represents the distance between the two objects. The variable k is the Coulomb's law constant, $9.0 \times 10^9 \ \mathrm{N \cdot m^2/C^2}$.

As with Newton's law of universal gravitation for the force of one planet on another, you're not going to be asked to calculate much with Coulomb's law. Rather, the questions will be qualitative and semiquantitative. For example, if the amount of charge A is doubled, what happens to the force of charge A on charge B? (It doubles.) Or, if you double the distance between two charges, what happens to the force of one on the other? (It is cut by one-fourth.)

Conservation of Charge

FACT: The total amount of charge in a system (or in the universe itself) is always the same.

Equal amounts of positive and negative charge can cancel out to make an object neutral, but those charges still exist on the object in the form of protons and electrons. Charge can be transferred from one object to another, for example by touching two charged metal spheres together or scuffing your feet on the carpet, but the total amount of charge stays the same.

Conservation of charge is usually discussed in conjunction with circuits and Kirchoff's junction rule, as discussed below.

Circuits

A circuit is any wire path that allows charge to flow. Technically, a current is defined as the flow of positive charge.[1] Under what conditions would this charge flow through a wire? This would occur when a coulomb of charge has a potential energy that's higher at one position in the wire than the other. We call this difference in potential energy per coulomb a "voltage," and a battery's job is to provide this voltage that allows current to flow. Current flows out of a battery from the positive side of the battery to the negative side.

Resistance Versus Resistivity

Resistance tells how difficult it is for charge to flow through something. Usually that "something" is a resistor, or a light bulb, or something that has a known or determinable resistance; usually the resistance of the wires connecting the "somethings" together is nothing, at least compared to the resistance of the things.

FACT: The resistance R of a wire of known dimensions is given by

$$R = \frac{\rho L}{A}$$

Sometimes, though, it's important to know how the properties of a wire affect its resistance. The longer the wire is, the more its resistance. The "wider" the wire is—that is, the bigger its cross-sectional area—the less its resistance. Two wires with the same shapes can have different resistances if they are made of different materials. Assuming the same shape, the wire with more resistance has a greater resistivity, represented by the variable ρ.

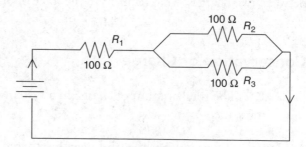

---

[1] No, not the flow of protons . . . okay, look, you really want to know? It's the flow of "holes in the electron sea." Understand? No? Neither do I. For the AP Exam, who cares. Current is the flow of positive charge, *capish*?

Example 1: The preceding circuit diagram contains a battery and three identical 100 Ω resistors.

Questions about circuits will occasionally ask for calculation: Find the voltage across each resistor and the current through each resistor. More often, though, you'll be asked qualitative questions, like which bulb takes the greatest current, or rank resistors from largest to smallest voltage across.

The Four Key Facts About Circuits

FACT: Series resistors each carry the same current, which is equal to the total current through the series combination.

FACT: The voltage across **series** resistors is different for each but adds to the total voltage across the series combination.

FACT: The voltage across **parallel** resistors is the same for each and equal to the total voltage across the parallel combination.

FACT: Parallel resistors each carry different currents, which add to the total current through the parallel combination.

Exam Tip from an AP Physics Veteran
Many first-year physics students are more comfortable making calculations with circuit problems than with explaining effects in words. If you are confused by a qualitative circuit question, try answering with a calculation: "Well, with a 150-V battery here's a calculation showing that I get 1 A of current in the circuit, but with a 75-V battery I only get 0.5 A. Thus, cutting the battery's voltage in half also cuts the current in half."

When you see a circuit, regardless of what questions about it are asked, it's worth making a *V-I-R* chart listing the voltage, current, and resistance for each resistor. Then the four facts above and Ohm's Law—the equation $V = IR$—can be used to find the missing value on any row of the chart. The easiest way to understand the *V-I-R* chart is to see it in action. Watch.

Start by sketching a chart and filling in known values. Right now, we know the resistance of each resistor. The voltage of the battery isn't given, so make it up. Try 100 V.[2]

---

[2] Why is it okay to make up a voltage that wasn't given? Say you're asked to rank the currents in the resistors. It doesn't matter how much current you calculate in each resistor, all that matters is whether R_1 or R_2 takes a bigger current. Any reasonable values will do for answering qualitative questions.

	V	I	R
R_1			100 Ω
R_2			100 Ω
R_3			100 Ω
total	100 V		

Next, simplify the circuit, collapsing sets of parallel and series resistors into their equivalent resistors.

FACT: The equivalent resistance of series resistors is the sum of all of the individual resistors. The equivalent resistance of parallel resistors is less than any individual resistor.

Strategy: When two *identical* resistors are connected in parallel, their equivalent resistance is half of either resistor. Here, then, the parallel combination of resistors has equivalent resistance 50 Ω. Even with nonidentical parallel resistors, you can usually estimate their equivalent resistance enough to do qualitative problems. If you need to make the calculation, the equivalent resistance of parallel resistors is given by $\frac{1}{R_{eq}} = \frac{1}{R_1} + \frac{1}{R_2}$.

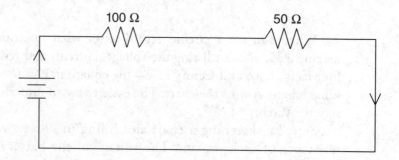

Since the other 100 Ω resistor is in series with the 50 Ω equivalent resistance, the equivalent resistance of the whole circuit is 150 Ω. We can put that equivalent resistance into the chart.

	V	I	R
R_1			100 Ω
R_2			100 Ω
R_3			100 Ω
total	100 V		150 Ω

Mistake: The V-I-R chart is not a magic square—it's merely a tool for organizing your calculations for a complicated circuit. You can *not* just add values up and down the columns.

Ah, progress: Two of the three entries in the "total" row are complete. Therefore, we can use Ohm's Law to calculate the total current in this circuit.

FACT: Ohm's law says that voltage across a circuit element equals that element's current times its resistance:

$$V = IR.$$

This equation can *only* be used across a single row in a *V-I-R* chart.

In the total row, (100 V) = I (150 Ω), giving a current in the circuit of $I = 0.67$ A.

Now what? We can't use Ohm's law because we don't have any rows missing just one entry. We go back to the Four Key Facts.

Resistor R_1 is in series with the battery; since current through series resistors is equal to the total current, R_1 must take the entire current flowing from the battery, all 0.67 A. Aha! Put 0.67 A in the chart for the current through R_1, and we can use Ohm's Law to calculate the voltage across R_1: that's 67 V. The chart appears as follows.

	V	I	R
R_1	67 V	0.67 A	100 Ω
R_2			100 Ω
R_3			100 Ω
total	100 V	0.67 A	150 Ω

Now to figure out R_2 and R_3. Look at their 50 Ω equivalent resistor in the redrawn diagram. It's in series with the 100 Ω resistor. Therefore, the 50 Ω resistor must add its voltage to 67 V to get the total voltage of 100 V and the voltage across the 50 Ω equivalent resistor is 33 V.

Then use the facts. Parallel resistors take the same voltage across each that is equal to the total voltage across the combination. Both R_2 and R_3 take 33 V across them. Fill in the chart.

	V	I	R
R_1	67 V	0.67 A	100 Ω
R_2	33 V		100 Ω
R_3	33 V		100 Ω
total	100 V	0.67 A	150 Ω

Now use Ohm's law across the rows for R_2 and R_3 to finish the chart.[3]

	V	I	R
R_1	67 V	0.67 A	100 Ω
R_2	33 V	0.33 A	100 Ω
R_3	33 V	0.33 A	100 Ω
total	100 V	0.67 A	150 Ω

This chart can now be used to answer any qualitative question. Sure, you should give more justification than just "look at my chart here." The chart will ensure you get the *right* answers, and that you have a clue about how to approach the qualitative questions.

For example, the exam might ask the following: Rank the voltage across each resistor from largest to smallest. Easy: $R_1 > R_2 = R_3$. Justify that with a verbal description of why you decided to use the calculations that you did: "Look at the simplification of the circuit to two series resistors. The voltage across these two series resistors must add to the voltage of the battery, but the current through them must be the same. By $V = IR$, the bigger resistor must take the larger voltage; this is R_1. Then R_2 and R_3 take equal voltage because they are in parallel with one another."

---

[3] Um, is that really right? The current through parallel resistors should add to the total current. But 0.33 A plus 0.33 A gives 0.66 A, and we said the total current was 0.67 A. This is *fine*. Expect that rounding in a *V-I-R* chart will not allow for ten-figure accuracy. Who cares—the whole point is generally to help answer conceptual questions, anyway.

Kirchoff's Laws: Conservation of Charge and Energy

FACT: Kirchoff's junction rule says that the current entering a wire junction equals the current leaving the junction.

This fact is a statement of conservation of charge: Since charge can't be created or destroyed, if 1 C of charge enters each second, the same amount each second must leave.

FACT: Kirchoff's loop rule says that the sum of voltage changes around a circuit loop is zero.

This fact is a statement of conservation of energy because voltage is a change in the electrical potential energy of 1 C of charge. A battery can raise the electrical potential energy of some charge that flows; a resistor will lower the potential energy of that charge.[4] But the sum of all these energy changes must be zero.

Look back at the Four Key Facts: These are just restatements of Kirchoff's laws and thus of conservation of energy and charge.

The junction rule applies to the facts about current. Series resistors take the same current through each, because there's no junction. The current through parallel resistors adds to the total because of the junction before and after the parallel combination.

The loop rule applies to the facts about voltage. The voltage across series resistors adds to the total voltage because the resistors can only drop the potential energy of 1 C of charge as much as the battery raised the charge's potential energy. The voltage across parallel resistors must be the same because Kirchoff's loop rule applies to all loops of the circuit. No matter which parallel path you look at, the sum of voltage changes must still be zero.

Power in a Circuit

Power is still defined as energy per second, just as it was in the chapter about energy. Resistors generally convert electrical energy to other forms of energy—the amount of power says how quickly that conversion occurs.

FACT: To determine the power dissipated by a resistor, use the equation

$$P = IV.$$

Of course, using Ohm's law, you can show that IV is equivalent to I^2R as well as $\dfrac{V^2}{R}$.

> **Exam Tip from an AP Physics Veteran**
> If an AP question asks about power, or equivalently, "the energy dissipated by a resistor per unit time," make a fourth column in your V-I-R chart. Use whichever of the power equations you can to calculate power.

---

[4] The potential energy can be converted by the resistor to internal energy of the resistor or the surrounding air, raising the temperature, as in a toaster oven; or could be converted to light, as in a lamp; or could even be converted to mechanical energy, as in an electric motor.

Power doesn't obey the Four Key Facts. The total power dissipated by a bunch of resistors is just the sum of the power dissipated by each, whether the resistors are in series, parallel, or whatever.

Circuits from an Experimental Point of View

When a real circuit is set up in the laboratory, it usually consists of more than just resistors—light bulbs and motors are common devices to hook to a battery, for example. For the purposes of computation, though, we can consider pretty much any electronic device to act like a resistor.

But what if your purpose is *not* computation? Often on the AP Exam, as in the laboratory, you are asked about observational and measurable effects. The most common questions involve the brightness of light bulbs and the measurement (not just computation) of current and voltage.

Brightness of a Bulb

The brightness of a bulb depends solely on the power dissipated by the bulb. (Remember, power is given by any of the equations I^2R, IV, or V^2/R). You can remember that from your own experience—when you go to the store to buy a light bulb, you don't ask for a 400-ohm bulb, but for a 100-watt bulb. And a 100-watt bulb is brighter than a 25-watt bulb. But be careful—a bulb's power can change depending on the current and voltage it's hooked up to. Consider this problem.

A light bulb is rated at 100 W in the United States, where the standard wall outlet voltage is 120 V. If this bulb were plugged in in Europe, where the standard wall outlet voltage is 240 V, which of the following would be true?

(A) The bulb would be one-quarter as bright.

(B) The bulb would be one-half as bright.

(C) The bulb's brightness would be the same.

(D) The bulb would be twice as bright.

(E) The bulb would be four times as bright.

Your first instinct might be to say that because brightness depends on power, the bulb is exactly as bright. But that's not right! The power of a bulb can change.

Under most operating conditions, the resistance of a lightbulb is a property of the bulb itself, and so it will not change much no matter to what the bulb is hooked.

That said, the resistance of a bulb can vary when the bulb's temperature is very cold (i.e., room temperature) or very hot. You can assume a bulb has constant resistance unless a problem clearly asks you to consider temperature variation.

Ammeters and Voltmeters

Ammeters measure current, and voltmeters measure voltage. This is pretty obvious, because current is measured in amps, voltage in volts. It is *not* necessarily obvious, though, how to connect these meters into a circuit.

Remind yourself of the properties of series and parallel resistors—voltage is the same for any resistors in parallel with each other. So if you're going to measure the voltage across a resistor, you must put the voltmeter in *parallel* with the resistor. In the following figure, the meter labeled V_2 measures the voltage across the 100 Ω resistor, while the meter labeled V_1 measures the potential difference between points A and B (which is also the voltage across R_1).

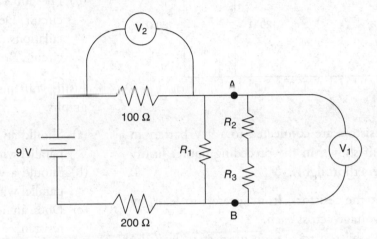

Measuring voltage with a voltmeter.

Current is the same for any resistors in *series* with one another. So, if you're going to measure the current through a resistor, the ammeter must be in series with that resistor. In the following figure, ammeter A_1 measures the current through resistor R_1, while ammeter A_2 measures the current through resistor R_2.

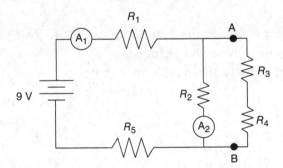

Measuring current with an ammeter.

As an exercise, ask yourself, is there a way to figure out the current in the other three resistors based only on the readings in these two ammeters? Answer is in the footnote.[5]

---

[5]The current through R_5 must be the same as through R_1, because both resistors carry whatever current came directly from the battery. The current through R_3 and R_4 can be determined from Kirchoff's junction rule: subtract the current in R_2 from the current in R_1 and that's what's left over for the right-hand branch of the circuit.

› Practice Problems

Note: Additional drills on circuit calculation can be found in Chapter 18.

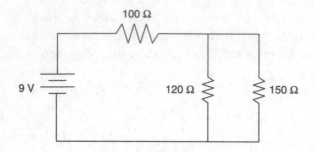

1. Three resistors are connected to a 9-V battery in the circuit shown in the preceding figure. Justify all answers thoroughly.

   (a) Rank the resistors, from greatest to least, by the voltage across each.

   (b) Rank the resistors, from greatest to least, by the current through each.

   (c) Rank the resistors, from greatest to least, by the power dissipated by each.

   (d) Is it possible to replace the 120 Ω resistor with a different resistor and change the voltage ranking? Answer thoroughly in a short paragraph.

   (e) Is it possible to replace the 120 Ω resistor with a different resistor and change the current ranking? Answer thoroughly in a short paragraph.

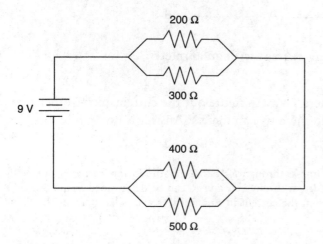

2. Four resistors are connected to a 9-V battery in the circuit in the diagram.

   (a) Calculate the equivalent resistance of the circuit.

   (b) Calculate the voltage across each resistor.

   (c) Calculate the current through each resistor.

   (d) The 500 Ω resistor is now removed from the circuit. Describe in words, without using calculations, what effect this would have on the circuit. Be specific about each resistor.

3. Justify your answers to the following in short paragraphs.

   (a) Should an ammeter be connected in series or parallel with the resistor it measures?

   (b) Should a voltmeter be connected in series or parallel with the resistor it measures?

   (c) Does an ideal ammeter have large or small resistance?

   (d) Does an ideal voltmeter have large or small resistance?

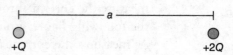

4. Two positive charges $+Q$ and $+2Q$ are separated by a distance a, as shown above.

   (a) Which is greater, the force of the $+Q$ charge on the $+2Q$ charge, or the force of the $+2Q$ charge on the $+Q$ charge?

   (b) In terms of given variables and fundamental constants, determine the magnitude and direction of the force of the $+2Q$ charge on the $+Q$ charge.

   (c) By what factor would the force calculated in (b) change if the distance between the charges were increased to $3a$?

   (d) Now the $+Q$ charge is replaced by a negative charge of the same magnitude, and the distance between the charges is returned to a. Describe how the magnitude and direction of the force exerted by each charge on the other will change from the original situation.

› Solutions to Practice Problems

1. (a) The circuit can be redrawn, simplifying the two parallel resistors to their equivalent resistance of 67 Ω. Then what's left is two resistors—100 Ω and 67 Ω—in series with the 9-V battery. Since series resistors take the same current through each, by $V = IR$ with I constant, the larger resistance takes the larger voltage. That's the 100 Ω resistor. Parallel resistors take the same voltage across each, equal to the voltage across the equivalent resistor. Ranking by voltage gives $V_{100\,\Omega} > V_{120\,\Omega} = V_{150\,\Omega}$.

   (b) Current only runs through a wire. All the current in the circuit must run through the 100 Ω resistor and then split at the junction with the parallel resistors. The parallel resistors each take the same voltage; by $V = IR$, with constant V, the smaller resistor takes the larger current. That's the 120 Ω resistor. The ranking by current is $I_{100\,\Omega} > I_{120\,\Omega} > I_{150\,\Omega}$.

   (c) Power is current times voltage. The 100 Ω resistor has the largest current *and* the largest voltage, so it has the greatest power. The other two resistors take the same voltage, but the 120 Ω resistor takes more current; thus, it has more power than the 150 Ω resistor. Ranking by power is $P_{100\,\Omega} > P_{120\,\Omega} > P_{150\,\Omega}$.

   (d) On one hand, it's not possible for the two parallel resistors to take different voltages, by definition (or by Kirchoff's loop rule, if you will). But it's totally possible for the parallel combination to take more voltage than the 100 Ω resistor. We need the equivalent resistance of the parallel combination to be greater than 100 Ω; then by $V = IR$ with I constant, the equivalent resistors would take higher voltage. Replace the 120 Ω resistor with, say, a 1,000 Ω resistor. Then the equivalent resistance of the parallel combination would be 130 Ω and would take more voltage than the 100 Ω resistor.

   (e) There's no way to avoid having the largest current go through the 100 Ω resistor; by Kirchoff's junction rule, it must take all the current in the circuit, while current splits between the other two resistors. Now, nothing says that the parallel resistor that comes first in the diagram has to take the larger current.

Replace the 120 Ω resistor with, say, a 200 Ω resistor—the voltage will still be the same for both parallel resistors, but now the 150 Ω resistor is the smaller one, taking the larger current.

2. (a) First simplify the two parallel combinations using $\frac{1}{R_{eq}} = \frac{1}{R_1} + \frac{1}{R_2}$. This gives 120 Ω for the top branch, and 220 Ω for the bottom branch. These two equivalent resistors are in series with one another, so their resistances add to the equivalent to give 340 Ω.

   (b) Each pair of parallel resistors takes the same voltage which is equal to that across the equivalent resistance of the pair. Treat this as series resistors of 120 Ω and 220 Ω connected to 9 V. The total current in the circuit is found with Ohm's law with the battery and the equivalent resistance of the whole circuit: $I = \frac{9\,V}{340\,\Omega} = 0.026$ A. (That's easier to understand as 26 mA.) The same 26 mA goes through each of the 120 Ω and 220 Ω resistors. Use Ohm's law with 0.026 A and each of these resistances to get voltages of 3.2 V and 5.8 V. The final answer:

   200 Ω resistor: 3.2 V.

   300 Ω resistor: 3.2 V

   400 Ω resistor: 5.8 V

   500 Ω resistor: 5.8 V

   (c) We know the voltage and resistance for each individual resistor. Ohm's law can be used to get the current through each by just dividing V/R. The answers are as follows (don't worry if you rounded slightly differently than I did):

   200 Ω resistor: 16 mA.

   300 Ω resistor: 11 mA

   400 Ω resistor: 15 mA

   500 Ω resistor: 12 mA

   (d) Start with the entire equivalent circuit. The 9-V battery is unchanged. The total resistance increases—the bottom branch previously had a resistance of 220 Ω but now is just the remaining 400 Ω. By $V = IR$ with constant V, a larger total resistance causes a smaller total current. And the 400 Ω resistor is now in series in the battery, and so it would take this entire current. The 200 Ω and 300 Ω resistors would still split the current, and in the same proportion; but the total current is smaller than before, so both will take smaller current now. As for voltage, the equivalent series circuit has more resistance in the second branch, so the 400 Ω resistor takes more of the 9 V than before, and the first parallel combination takes less voltage than before.

3. (a) Series resistors take the same current as each other, which is equal to the total—that's Kirchoff's junction rule. An ammeter measures current; it should be connected in series so that all current that passes through the resistor also passes through the ammeter.

(b) Parallel resistors take the same voltage as each other, which is equal to the total—that's Kirchoff's loop rule. A voltmeter measures voltage; it should be connected in parallel so that it also takes the same voltage as the resistor it's measuring.

(c) An ammeter is in series with a resistor. If the ammeter has a large resistance, then it also takes a large voltage, leaving less voltage to go across the resistor, and affecting the circuit. If the ammeter has a very *small* resistance, then it takes hardly any of the total voltage, leaving the resistor to have the same voltage and current as in a circuit without the ammeter.

(d) A voltmeter is in parallel with a resistor. If the voltmeter has a small resistance, then more current will choose the parallel path with the voltmeter than with the resistor, affecting the circuit. But if the voltmeter has a very *large* resistance, then almost none of the current would take the parallel path with the voltmeter in it, leaving the resistor to have the same voltage and current as in a circuit without the voltmeter.

4. (a) These forces are the same—Newton's third law applies to all forces, even electrical forces.

(b) Coulomb's law is the relevant equation. Using the variables as given, the force of one charge on the other is $k\dfrac{(Q)(2Q)}{a^2}$. You can rearrange this to $\dfrac{2kQ^2}{a^2}$ if you want, but it's not mandatory for AP-style credit, but notice that you should *not* be including a plus sign anywhere, or a minus sign if the charges were negative. The magnitude of a force should not have any signs on it. The direction is repulsive, so the $+2Q$ charge pushes the $+Q$ charge to the left.

(c) The distance between charges is in the denominator, so increasing the distance decreases the force. The distance term is squared, so that multiplying a by 3 multiplies the whole denominator by $3^2 = 9$. Therefore, the magnitude of the force is reduced by a factor of 9.

(d) Now the force between charges is attractive, meaning that the $-Q$ charge will be pulled to the right. The magnitude of the force will not change from Part (b), since the amount of charge of each item and the distance between charges will not change.

› Rapid Review

- Coulomb's law is an equation for the force exerted by one electrical charge on another:

$$F = k\frac{Q_1 Q_2}{d^2}$$

- The total amount of charge in a system (or in the universe itself) is always the same.
- The resistance R of a wire of known dimensions is given by $R = \dfrac{\rho L}{A}$.

- **Series** resistors each carry the same current, which is equal to the total current through the series combination.

- The voltage across **series** resistors is different for each but adds to the total voltage across the series combination.

- The voltage across **parallel** resistors is the same for each and equal to the total voltage across the parallel combination.

- **Parallel** resistors each carry different currents, which add to the total current through the parallel combination.

- The equivalent resistance of series resistors is the sum of all of the individual resistors. The equivalent resistance of parallel resistors is less than any individual resistor (and, if you have to do the calculation, is given by $\frac{1}{R_{eq}} = \frac{1}{R_1} + \frac{1}{R_2}$).

- Ohm's law says that voltage across a circuit element equals that element's current times its resistance: $V = IR$. This equation can *only* be used across a single row in a *V-I-R* chart.

- Kirchoff's junction rule says that the current entering a wire junction equals the current leaving the junction.

- Kirchoff's loop rule says that the sum of voltage changes around a circuit loop is zero.

- To determine the power dissipated by a resistor, use the equation $P = IV$.

CHAPTER 17

Waves and Simple Harmonic Motion

IN THIS CHAPTER

Summary: This chapter introduces basic properties of waves, especially of sound waves. You'll define wave speed, frequency, and wavelength, and relate them through $v = \lambda f$. Although a wave moves through a material, the pieces of the material themselves do not move. Rather, they tend to oscillate in simple harmonic motion. This chapter describes exactly what that means.

Definitions

- ✪ The **period** is the time for one cycle of simple harmonic motion, or the time for a full wavelength to pass a position.
- ✪ The **frequency** is the number of cycles, or the number of wavelengths passing a position, in one second.
- ✪ The unit of frequency is the Hz, which means "per second."
- ✪ The **amplitude** is the distance from the midpoint of a wave to its crest, or the distance from the midpoint of simple harmonic motion to the maximum displacement.
- ✪ The **wavelength** is measured from peak to peak, or between any successive identical points on a wave.
- ✪ The **spring constant** k, measured in units of newtons per meter (N/m), is related to the stiffness of a spring.
- ✪ A **restoring force** is any force that always pushes an object toward an equilibrium position.

⊗ **Nodes** are the stationary points on a standing wave.
⊗ **Antinodes** are the positions on standing waves with the largest amplitudes.

You're probably most familiar with waves on the surface of a lake or pond. Those are transverse waves, which look like the waves on the machine in the following figure. The wavelength of the wave can be measured with a ruler from peak to peak.

Simple Harmonic Motion

Simple harmonic motion refers to a back-and-forth oscillation whose position-time graph looks like a sine function. The typical examples are a mass vibrating on a spring, and a pendulum.

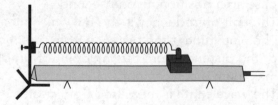

Example: A cart of mass 0.5 kg is attached to a spring of spring constant 30 N/m on a frictionless air track, as shown. The cart is stretched 10 cm from the equilibrium position and released from rest.

FACT: The period of a mass on a spring in simple harmonic motion is given by

$$2\pi \frac{\sqrt{m}}{\sqrt{k}}.$$

Of course, the AP Exam will not likely ask, "What is the period of this oscillation?" Rather, it might ask for the specific effect that doubling the mass of the cart would have on the period. Since the m term is in the numerator of the period equation, a bigger mass means a larger (longer) period of oscillation. Since the m is under a square root, doubling the mass multiplies the period by the square root of 2.

What if the amplitude of the motion were doubled? How would that affect the period? Since you don't see an A in the equation for the period, the period would not change. This is a general result for all simple harmonic motion and wave problems: The amplitude does not affect the period.[1]

FACT: The frequency and period are inverses of one another.

Once you know by being told or by doing the calculation that the period of this mass on a spring is 0.81 s, you can use your calculator to do 1 divided by 0.81 s, giving a frequency of 1.2 s.

FACT: The amount of restoring force exerted by a spring is given by

$$F = kx.$$

The force of the spring on the cart is therefore greatest when the spring is most stretched, but zero at the equilibrium position. And since $a = F_{\text{net}}/m$, the acceleration likewise changes from lots at the endpoints, to nothing at the middle.

This means that you cannot use kinematics equations with harmonic motion. A kinematics approach requires constant acceleration. Instead, when a problem asks for the speed of a cart somewhere, use conservation of energy.

FACT: The spring potential energy is given by

$$PE_{\text{spring}} = \frac{1}{2}kx^2.$$

---

[1]There are exceptions for simple and physical pendulums once the amplitude reaches large enough values, but the AP will not likely ask much about these situations.

The energy stored by the spring is thus largest at the endpoints and zero at the equilibrium position. There, the spring energy is completely converted to the kinetic energy of the cart. Where is the cart's speed greatest, then? At the equilibrium position, of course, because kinetic energy is $\frac{1}{2}mv^2$—largest kinetic energy means largest speed.

To calculate the value of the maximum speed, write out the energy conversion from the endpoint of the motion to the midpoint of the motion: spring potential energy is converted to kinetic energy. Translated into equations, you get $\frac{1}{2}kx^2 = \frac{1}{2}mv^2$. Plug in values, and solve for the speed. Here, you get 0.77 m/s (i.e., 77 cm/s).[2]

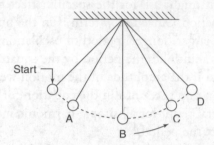

Example 2: The pendulum shown in the preceding figure is released from rest at the start position. It oscillates through the labeled positions A, B, C, and D.

Treat a pendulum pretty much the same way as a spring. It's still in harmonic motion; it still requires an energy approach, not a kinematics approach, to determine its speed at any position.

FACT: The period of a pendulum is given by

$$2\pi \frac{\sqrt{L}}{\sqrt{g}}.$$

As always, it's unlikely you're going to plug in numbers to calculate a period. Among the gazillion possibilities, you might well be asked to rank the listed positions in terms of some quantity or other. Here are some ideas:

Rank the lettered positions from greatest to least by the bob's gravitational potential energy. Gravitational potential energy is *mgh*; the bob always has the same mass and *g* can't change, so the highest vertical height has the greatest gravitational potential energy. Ranking: D > C = A > B.

Rank the lettered positions from greatest to least by the bob's total mechanical energy. Total mechanical energy means the sum of potential and kinetic energies. Here, with

---

[2]Don't forget to convert the maximum distance from equilibrium to 0.1 meters before plugging into the equation.

no nonconservative force like friction acting, and no internal structure to allow for internal energy, the total mechanical energy doesn't change. The ranking is as follows: (A = B = C = D).

Rank the lettered positions from greatest to least by the bob's speed. Since gravitational potential energy is converted to kinetic energy, the bob moves fastest when the gravitational potential energy is smallest. Ranking: B > C = A > D.

The gravitational field at the surface of Jupiter is 26 N/kg and on the surface of the Moon, 1.6 N/kg. Rank this pendulum's period near these two planets and earth. Since g is in the denominator of the period equation, the lowest gravitational field will have the greatest period; so $T_{Moon} > T_{Earth} > T_{Jupiter}$. The ranking by frequency would be just the opposite—because frequency is the inverse of period, a bigger g leads to a smaller period but a bigger frequency.

Waves

The AP Physics 1 Exam covers only "mechanical" waves, such as sound, or waves on the surface of the ocean. Light waves (i.e., electromagnetic waves) are not covered in detail.

FACT: Whenever the motion of a material is at right angles to the direction in which the wave travels, the wave is a **transverse wave.**

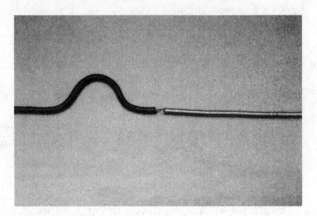

Example 3: A wave pulse travels to the right through a spring and then extends into a second spring in the preceding figure. The speed of the waves is faster on the right-hand spring.

This is a transverse wave pulse—the coils of the spring travel up and down the page, while the wave itself moves to the right.

FACT: The energy carried by a wave depends on the wave's amplitude.

A good AP-style question might ask you to resketch the diagram so that a pulse of about the same wavelength carries more energy. Make the amplitude—the maximum displacement of the coil above the resting position—bigger, then, because amplitude is related to energy carried by a wave. Keep the pulse about the same length.

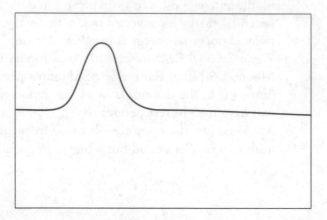

FACT: When a wave changes materials, its frequency remains the same.

When this waves moves into the new spring, the wave speeds up, but the frequency remains the same. By $v = \lambda f$, the wavelengths will also increase, because multiplying the same frequency by the wavelength has to give a bigger value for v. The wave will look wider, then, in the new spring.

FACT: When a material vibrates parallel to the direction of the wave, the wave is a **longitudinal wave.**

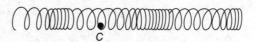

Example 4: A wave travels through a spring. A picture of the spring is shown in the preceding figure, with point C labeled.

This wave is a longitudinal wave—the disturbance is traveling through the spring, so it is traveling right or left. The coils of the spring itself are spread out and compressed. Since the motion of the spring's coils is also left-right, parallel to the way the disturbance is traveling, this is a longitudinal wave.

Try drawing a wavelength on the picture. A wavelength is defined as the distance between two identical positions on the wave. From position C to the next spot where the coils are all stretched out is one wavelength.

Superposition and Interference

When two waves collide, they don't bounce or stick like objects do. Rather, the waves interfere—they form one single wave for just a moment, and then the waves continue on their merry way.

 **FACT:** In **constructive interference** the crest of one wave overlaps the crest of another. The result is a wave of increased amplitude.

Two wave pulses about to interfere constructively.

The two wave pulses on a string moving toward each other in the preceding figure will interfere constructively, since their amplitudes are on the same side of the string's resting position. When the pulses meet, the wave will look like the dark line in the following figure.

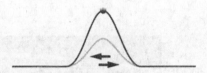

Two wave pulses interfering constructively.

Then the waves continue on in the direction they were originally traveling (see the following figure).

Two wave pulses after interfering constructively.

 **FACT:** In **destructive interference** the crest of one wave overlaps the trough of another. The result is a wave of reduced amplitude.

The same principle applies to waves with amplitudes on opposite sides of the string's resting position, producing destructive interference.

If we send the two wave pulses in the following figure toward each other, they will interfere destructively (next figure following) and then continue along their ways (last figure following).

Two wave pulses about to interfere destructively.

Two wave pulses interfering destructively.

Two wave pulses after interfering destructively.

Standing Waves

A **standing wave** by definition is a wave that appears to stay in one place. In some positions, the strings vibrate with large amplitude—these are called **antinodes**. In other positions, the strings don't vibrate at all—these are called **nodes**.

The reason that a standing wave exists revolves around interference. Waves are traveling back and forth on the string, reflecting off of each end and interfering with each other all willy-nilly. The net effect of all this interference is a pattern of nodes and antinodes.

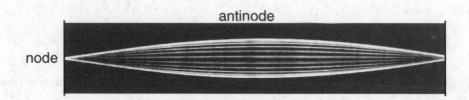

Example 5: A guitar string of length 1 m is plucked.

Generally when you pluck a string, you produce a standing wave of the longest possible wavelength, and thus the smallest possible frequency. The string will look like the preceding picture if you watch it carefully.

FACT: The wavelength of a standing wave is twice the node-to-node distance.

In this case, the node-to-node distance is equal to the length of the string: 1 m. So the wavelength here is 2 m.

But what if you put your finger very lightly on the middle of the string, forcing a node to exist there? Then you create a harmonic, like the one shown in the following figure. Now the wavelength of this standing wave is 1 m.

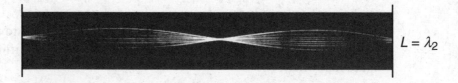

$L = \lambda_2$

And you can put not just one or two, but any whole number of antinodes on a string. (See the following figure.)

FACT: For a string fixed at both ends, or for a pipe open at both ends, the smallest frequency of standing waves is given by

$$f_1 = \frac{v}{2L}$$

where v is the speed of the waves on the string or in the pipe. Other harmonic frequencies for this string or pipe must be whole-number multiples of the fundamental frequency.

When we're dealing with an open pipe, the speed v is generally the speed of sound in air, or about 340 m/s. For a string, the wave speed depends on the tension in the string and the mass of 1 meter worth of string—usually you're talking a few hundred meters per second (m/s), but that can vary.

So if this guitar string produces a fundamental frequency of, say, 100 Hz, then the harmonics that can be played are 200 Hz, 300 Hz, 400 Hz, etc. This string cannot play a frequency of 150 Hz or 370 Hz, at least unless the length of the string or the tension in the string is changed.

FACT: The pitch of a musical note depends on the sound wave's frequency; the loudness of a note depends on the sound wave's amplitude.

A guitar can generally play any note in a musical scale. But how can that be, if the harmonics are restricted to multiples of the fundamental frequency? Producing harmonics is generally not the way to play a guitar.[3] Rather, the frets on the neck are used to shorten or lengthen the string; since the speed of waves on the string is unchanged, shortening the string lowers the L in $f_1 = \frac{v}{2L}$. The fundamental frequency of the shorter string will be higher, and thus the note played will be higher pitched.

The guitar can be tuned by tightening or loosening the string. A tighter string, for example, will produce a standing wave with higher wave speed. For the same length of string and thus the same wavelength, the frequency will be higher by $v = \lambda f$. A higher frequency means a higher-pitched note.

Closed-Ended Pipe

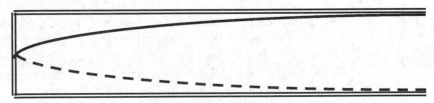

Example 6: A 1-m-long PVC pipe is covered at one end and open at the other.

When a pipe is open at one end and closed at the other, the standing wave in this pipe must have a node at one end and an antinode at the other. The wave with the longest possible

---

[3]Harmonics (other than the fundamental) are not the notes you hear. The *overtones* (the harmonics higher than the fundamental) are what give the note its *quality*, which makes a guitar sound like a guitar or a violin sound like a violin. After all, the strings themselves are about the same, so did you ever wonder how it was you are able to tell the difference between the sounds of stringed instruments?

wavelength is shown in the preceding figure. The wavelength of a standing wave is always twice the node-to-node distance. But the wave is so long that we don't even see a second node. It turns out that the wavelength of this wave is 4 m (four times the length of the pipe).

FACT: For a pipe closed at one end, or for a string fixed at one end but free at the other, the smallest frequency of standing waves is given by

$$f_1 = \frac{v}{4L}$$

where v is the speed of the waves on the string or in the pipe. Other harmonic frequencies for this string or pipe must be *odd* multiples of the fundamental frequency.

The speed of waves in this pipe is the speed of sound, or about 340 m/s. The fundamental frequency is 85 Hz. The other frequencies that this pipe can produce are only the *odd* multiples of 85 Hz: 255 Hz, 425 Hz, etc.

Beats

FACT: Beats are rhythmic interference that occurs when two notes of unequal but close frequencies are played.

If you have a couple of tuning forks of similar—but not identical—frequency to play with, or if you have a couple of tone generators at your disposal, you might enjoy generating some beats of your own. They make a wonderful "wa-wa" sound, which is due to a periodic increase and decrease in intensity, or loudness. The frequency of the "wa-wa" is equal to the *difference* between the two frequency generators.

Doppler Effect

Whenever a fire engine or ambulance races by you with its sirens blaring, you experience the **Doppler effect**. Similarly, if you enjoy watching auto racing, that "Neeee-yeeeer" you hear as the cars scream past the TV camera is also attributable to the Doppler effect.

FACT: The Doppler effect is the apparent change in a wave's frequency that you observe whenever the source of the wave is moving toward or away from you.

To understand the Doppler effect, let's look at what happens as a fire truck travels past you (see the following two figures).

When the fire truck moves toward you, the sound waves get squished together, increasing the frequency you hear.

As the fire truck moves away from you, the sound waves spead apart, and you hear a lower frequency.

As the fire truck moves toward you, its sirens are emitting sound waves. Let's say that the sirens emit one wave pulse when the truck is 50 meters away from you. It then emits another pulse when the truck is 49.99 meters away from you. And so on. Because the truck keeps moving toward you as it emits sound waves, it appears to you that these waves are getting scrunched together.

Then, once the truck passes you and begins to move away from you, it appears as if the waves are stretched out.

Now, imagine that you could record the instant that each sound wave hits you. When the truck is moving toward you, you would observe that the time between when one wave hits and when the next wave hits is very small. However, when the truck is moving away from you, you would have to wait a while between when one wave hits you and when the next one does. In other words, when the truck is moving toward you, you register that the sirens are making a higher frequency noise; and when the truck is moving away, you register that the sirens are making a lower frequency noise.

That's all you really need to know about the Doppler effect. Just remember, the effect is rather small—at normal speeds, the frequency of, say, a 200 Hz note will only change by a few tens of Hz, not hundreds of Hz.

› Practice Problems

Note: Additional drills regarding simple harmonic motion graphs are included in Chapter 18.

1. In a pipe closed at one end and filled with air, a 384-Hz tuning fork resonates when the pipe is 22-cm long; this tuning fork does not resonate for any smaller pipes.

   (a) State three other lengths at which this pipe will resonate with the 384-Hz tuning fork.

   (b) The end of the pipe that was closed is now opened, so that the pipe is open at both ends. Describe any changes in the lengths of pipe that will resonate with the 384-Hz tuning fork.

   (c) The air in the closed pipe is replaced with helium. Describe an experiment that would use the pipe to determine the speed of sound in helium.

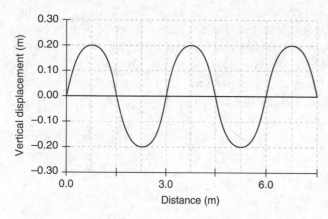

2. The wave shown in the previous figure travels in a material in which its speed is 30 m/s.

   (a) What is the wavelength of this wave?

   (b) Calculate the frequency of this wave.

   (c) On the diagram, draw a different wave that has a larger frequency but carries less energy than the one pictured.

3. Consider the following questions about electromagnetic and mechanical waves. Justify your answer to each.

   (a) Which of the following types of wave can be transmitted through space, where there is no air? Choose all that apply. Justify your answer briefly.
      (i) Visible light
      (ii) Radio waves
      (iii) Gamma rays
      (iv) Sound waves

   (b) (multiple choice) A tuning fork vibrating in air produces sound waves. These waves are best classified as
      (A) Transverse, because the air molecules are vibrating parallel to the direction of wave motion
      (B) Transverse, because the air molecules are vibrating perpendicular to the direction of wave motion
      (C) Longitudinal, because the air molecules are vibrating parallel to the direction of wave motion
      (D) Longitudinal, because the air molecules are vibrating perpendicular to the direction of wave motion

   (c) (multiple choice) Radio waves and gamma rays traveling in space have the same
      (A) Frequency
      (B) Wavelength
      (C) Period
      (D) Speed

   (d) Which type of wave exhibits the Doppler effect? Choose all that apply. Justify your answer briefly.
      (i) Visible light
      (ii) Radio waves
      (iii) Gamma rays
      (iv) Sound waves

4. The driver of a car blows the horn as the car approaches you.

   (a) Compared to the horn's pitch heard by the driver, will the pitch observed by you be higher, lower, or the same?

   (b) The car passes you, while the driver continues to blow the horn. After the car passes, you notice that the horn doesn't sound as loud as it did when it was near you. Is this observation a result of the Doppler effect?

   (c) The car recedes from you after passing you, still producing sound waves from the horn. Discuss how the amplitude, period, and frequency of the sound waves that you would measure compare to the amplitude, period, and frequency of the sound waves that the driver would measure.

5. The period of a mass-on-a-spring system is doubled, while still using the same spring.

(a) By what factor does the frequency of the mass-on-a-spring system increase or decrease?

(b) By what factor does the mass attached to the spring increase or decrease?

❯ Solutions to Practice Problems

1. (a) Because 22 cm is the shortest length of pipe that resonates, 384 Hz is the fundamental frequency, the one that produces waves without any nodes inside the pipe. There must be an antinode at one end and a node at the other (because it's closed at one end and open at the other). As the pipe length is increased, the wavelength of the sound wave doesn't change, because the frequency of the tuning fork and the speed of sound don't change. The pipe will next resonate when once again there is an antinode at one end and a node at the other. (This time, though, there will be another node inside somewhere, too.) Since the antinode-to-node distance was 22 cm, we need to add that distance twice to get another full "hump" of a standing wave inside the pipe. Add 44 cm to the pipe to get resonance at a pipe length of 66 cm; add another 44 cm to get resonance at 110 cm; and add another 44 cm to get resonance at 154 cm.

(b) Now only standing waves with nodes at both ends of the pipe will resonate. The wave of 22 cm has a node at one end and an antinode at the other—it will no longer resonate. But 44 cm is the node-to-node distance, so 44 cm will resonate. Whereas 66 cm used to resonate in the closed pipe, it will not in the open pipe because an antinode is at one end. Instead, any multiple of 44 cm will resonate because adding 44 cm adds a full node-to-node distance, ensuring a node at each end.

(c) We know the frequency of the tuning fork. To use the equation $v = \lambda f$, we need the wavelength of the wave as well. Play the tuning fork and shorten the pipe until once again we find the shortest pipe length that resonates with the tuning fork. That's the fundamental, with an antinode at one end, a node at the other, and no nodes in between. The wavelength of the sound wave is four times this shortest resonating pipe length.

2. (a) The wavelength is measured from peak to peak or trough to trough. That's 3.0 m.

(b) Use $v = \lambda f \ldots$ (30 m/s) = (3.0 m)$f \ldots f$ = 10 Hz.

(c) The energy carried by a wave depends on the wave's amplitude; this wave should have a smaller amplitude. Since the wave speed doesn't change, a bigger frequency means a smaller wavelength by $v = \lambda f$.

3. (a) Correct answers: (i), (ii), and (iii). Only electromagnetic waves can be transmitted through a vacuum. That includes gamma rays, visible light, and radio waves, which are all parts of the electromagnetic spectrum. Sound waves require a material to be transmitted.

(b) Correct answer: (C). Sound waves are longitudinal, by definition. Also by definition, "longitudinal" means that the particles of the material are vibrating parallel to the direction that the wave travels.

(c) Correct answer: (D). The speed of light—or any electromagnetic wave—in a vacuum is 300 million m/s. Radio and gamma rays have different frequencies, and thus different wavelengths and periods.

(d) Correct answers: (i), (ii), (iii), (iv). All waves exhibit the Doppler effect. When the source of the wave moves toward an observer, the observer observes waves of higher frequency. For sound waves, this means you'd hear a higher pitch; for visible light, this means you'd see a "blue shift." Gamma and radio waves would also be observed to have a higher frequency.

4. (a) By definition, when a source of waves approaches an observer, the observer observes waves of a higher frequency. Pitch of a sound is related to the sound wave's frequency.

   (b) The Doppler effect says nothing about loudness, only about frequency. The horn sounds less loud because the energy created by the wave source spreads out over a larger and larger space as the wave gets farther from the source. Since the energy carried by a wave is related to its amplitude, and since amplitude is related to loudness for a sound wave, you hear a softer noise.

   (c) The horn won't seem as loud (as discussed in [b]), so the amplitude is smaller than what the driver hears. The frequency you hear will be lower based on the Doppler effect. Since period is the inverse of frequency, a smaller frequency means a larger period.

5. (a) Frequency is the inverse of the period. When the period doubles, the frequency is cut in half.

   (b) The relevant equation is $T = 2\pi \frac{\sqrt{m}}{\sqrt{k}}$. The spring constant doesn't change because it's the same spring. Since the mass term is in the numerator and square rooted, the mass should quadruple. Then, square rooting the factor of four increases the whole fraction by a factor of two.

› Rapid Review

- The period of a mass on a spring in simple harmonic motion is given by $2\pi \frac{\sqrt{m}}{\sqrt{k}}$.
- The frequency and period are inverses of one another.
- The amount of restoring force exerted by a spring is given by $F = kx$.
- The spring potential energy is given by $PE_{\text{spring}} = \frac{1}{2}kx^2$.
- The period of a pendulum is given by $2\pi \frac{\sqrt{L}}{\sqrt{g}}$.
- Whenever the motion of a material is at right angles to the direction in which the wave travels, the wave is a **transverse wave.**
- The energy carried by a wave depends on the wave's amplitude.
- When a wave changes materials, its frequency remains the same.
- When a material vibrates parallel to the direction of the wave, the wave is a **longitudinal wave.**
- In **constructive interference** the crest of one wave overlaps the crest of another. The result is a wave of increased amplitude.
- In **destructive interference** the crest of one wave overlaps the trough of another. The result is a wave of reduced amplitude.
- The wavelength of a standing wave is twice the node-to-node distance.
- For a string fixed at both ends, or for a pipe open at both ends, the smallest frequency of standing waves is given by $f_1 = \frac{v}{2L}$, where v is the speed of the waves on the string or in the pipe. Other harmonic frequencies for this string or pipe must be whole-number multiples of the fundamental frequency.
- The pitch of a musical note depends on the sound wave's frequency; the loudness of a note depends on the sound wave's amplitude.

- For a pipe closed at one end, or for a string fixed at one end but free at the other, the smallest frequency of standing waves is given by $f_1 = \frac{v}{4L}$, where v is the speed of the waves on the string or in the pipe. Other harmonic frequencies for this string or pipe must be *odd* multiples of the fundamental frequency.

- Beats are rhythmic interference that occurs when two notes of unequal but close frequencies are played.

- The **Doppler effect** is the apparent change in a wave's frequency that you observe whenever the source of the wave is moving toward or away from you.

Extra Drills on Difficult but Frequently Tested Topics

IN THIS CHAPTER

Summary: Included in this chapter are problems providing extra practice on frequently tested topics that students often find difficult:

- ✪ Springs and graphs
- ✪ Tension
- ✪ Inclined planes
- ✪ Motion graphs
- ✪ Simple circuits

If you have any extra time, spend it sharpening your AP Physics 1 skills working out these problems. Following each question set are detailed explanations that take you step-by-step to the solution.

How to Use This Chapter

Practice problems and tests cannot possibly cover every situation that you may be asked to understand in physics. However, some categories of topics come up again and again, so much so that they might be worth some extra review. And that's exactly what this chapter is for—to give you a focused, intensive review of a few of the most essential physics topics.

We call them "drills" for a reason. They are designed to be skill-building exercises, and as such, they stress repetition and technique. Working through these exercises might remind you of playing scales if you're a musician or of running laps around the field if you're an athlete. Not much fun, maybe a little tedious, but very helpful in the long run.

The questions in each drill are all solved essentially the same way. *Don't* just do one problem after the other . . . rather, do a couple, check to see that your answers are right,[1] and then, half an hour or a few days later, do a few more, just to remind yourself of the techniques involved.

Springs and Graphs

The Drill

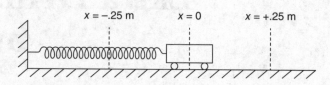

A 0.50-kg lab cart on a frictionless surface is attached to a spring, as shown in the preceding figure. The rightward direction is considered positive. The spring is neither stretched nor compressed at position $x = 0$. The cart is released from rest at the position $x = +0.25$ m at time $t = 0$.

1. On the axes below, sketch a graph of the kinetic energy of the cart as a function of position x.

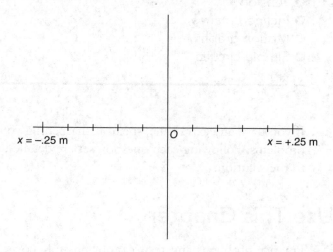

---

[1]For numerical answers, it's okay if you're off by a significant figure or so.

2. On the axes below, sketch a graph of the total mechanical energy of the cart-spring system as a function of position x.

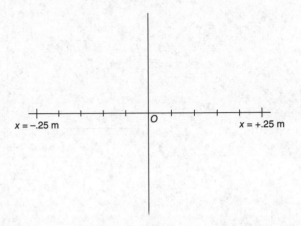

$x = -.25$ m O $x = +.25$ m

3. On the axes below, sketch a graph of the speed of the cart-spring system as a function of position x.

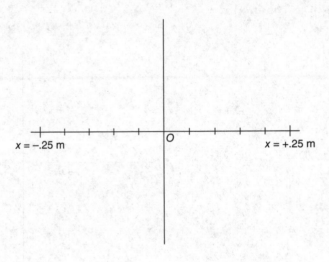

$x = -.25$ m O $x = +.25$ m

4. On the axes below, sketch a graph of the force applied by the spring on the cart as a function of position x.

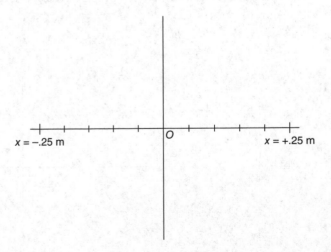

$x = -.25$ m O $x = +.25$ m

5. On the axes below, sketch a graph of the force applied by the cart on the spring as a function of position *x*.

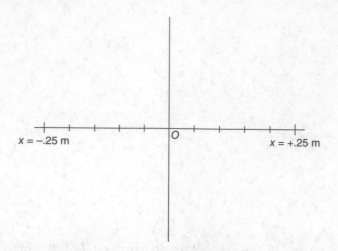

6. On the axes below, sketch a graph of the spring constant of the spring as a function of position *x*.

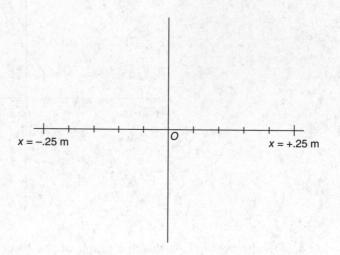

7. On the axes below, sketch a graph of the acceleration of the cart-spring system as a function of position *x*.

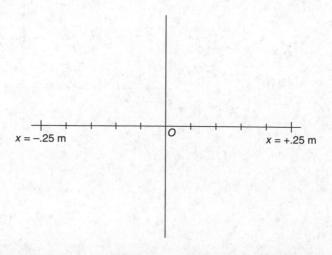

8. On the axes below, sketch a graph of the potential energy of the cart-spring system as a function of position x.

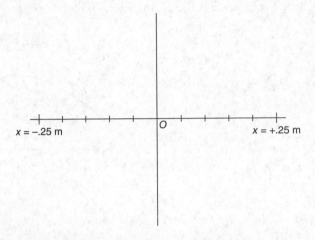

x = −.25 m   O   x = +.25 m

Answers with Explanations

In this section, the figure is shown first, followed by the explanation.

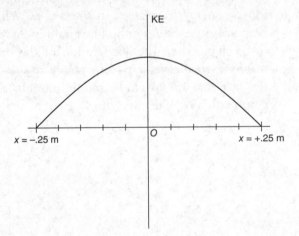

KE

x = −.25 m   O   x = +.25 m

1. Kinetic energy is zero at the endpoints, where the cart comes briefly to rest. At the midpoint, the cart-spring system's potential energy ($\frac{1}{2}kx^2$) is zero since $x = 0$. All the energy is kinetic at the midpoint. The graph is curved because the potential energy graph is curved; the kinetic and potential energy must add to the same value, since friction is negligible.

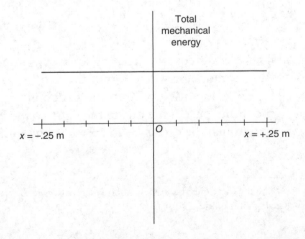

Total
mechanical
energy

x = −.25 m   O   x = +.25 m

2. Total mechanical energy of a system with no nonconservative forces doing work is conserved. Friction is negligible; gravity and the normal force do no work. The only relevant force is the spring force, which is conservative. Mechanical energy cannot change.

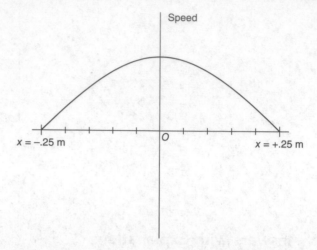

3. The cart comes briefly to rest at both ends, so speed is zero there. At the midpoint, kinetic energy is greatest, so speed is as well: $KE = \frac{1}{2}mv^2$. Speed has no direction, so graph positive values only. The graph is curved—you can know that because the spring force changes, so the acceleration of the cart is not constant. Constant acceleration means a straight velocity-time graph, so this graph must be curved.[2] Or, you can know the fact that in simple harmonic motion, a position-time graph and a velocity-time graph will look like curvy sine functions.

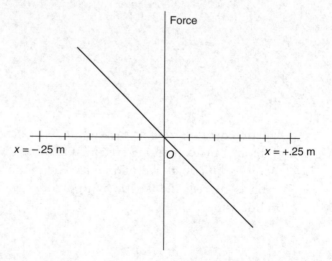

4. The relevant equation relating force of a spring to displacement of the spring is $F = kx$. At the midpoint where $x = 0$, the net force is zero as well. The amount of force gets bigger as the distance from the midpoint gets bigger. When the cart is far to the left, the spring pushes the cart to the right; so the force is positive when the distance is negative. When the cart is far to the right, the spring pushes the cart to the left; so the force

---

[2]And the curve must be steepest at the endpoints, where the net force $F = kx$ is greatest; that's where the speed must be changing most quickly.

is negative when the distance is positive. The graph is straight because the x is neither squared nor square rooted, but just to the first power.

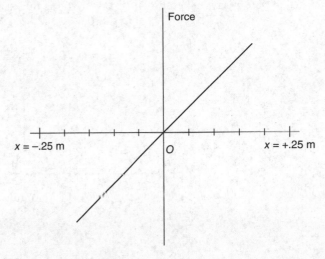

5. By Newton's third law, the force of the cart on the spring is equal in magnitude and opposite in direction to the force of the spring on the cart. Just flip the direction of the answer to Question 4.

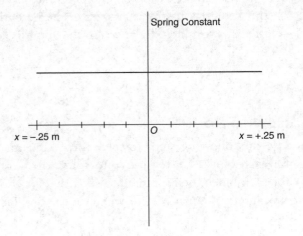

6. The spring constant of a spring is a property of the spring itself. Since the spring isn't replaced while the cart oscillates, the spring constant cannot change.

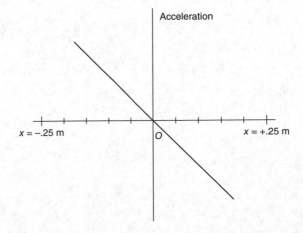

7. By Newton's second law, the net force on the cart is equal to *ma*. The acceleration graph should look exactly like the force of the spring on the cart graph. If we cared about the actual values on the graph, we'd divide the force by the cart's mass at each position. The problem says "sketch"—just the shape is necessary.

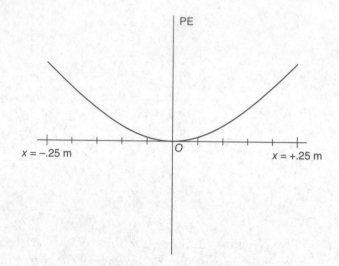

8. The relevant equation for the potential energy due to a spring is $PE = \frac{1}{2}kx^2$. At $x = 0$, the potential energy is also zero. The potential energy is largest at the endpoints. And the equation includes an x^2, so the graph is curved.

Tension

How to Do It

Use the following steps to solve these kinds of problems: (1) Draw a free-body diagram for each block; (2) resolve vectors into their components; (3) write Newton's second law for each block, being careful to stick to your choice of positive direction; and (4) solve the simultaneous equations for whatever the problem asks for.

The Drill

In the diagrams below, assume all pulleys and ropes are massless, and use the following variable definitions:

$$F = 10 \text{ N}$$
$$M = 1.0 \text{ kg}$$
$$\mu = 0.2$$

Find the tension in each rope and the acceleration of the set of masses.
(For a greater challenge, solve in terms of *F*, *M*, and μ instead of plugging in values.)

1. Frictionless

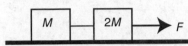

2. Frictionless

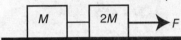

3. Frictionless

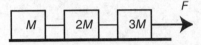

4. Coefficient of Friction μ

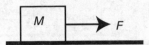

5.

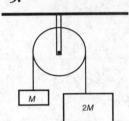

6.

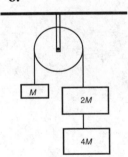

7. Frictionless

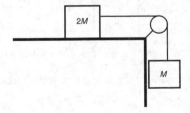

8. Frictionless

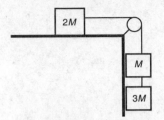

9. Frictionless

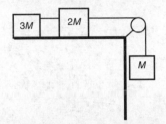

10. Coefficient of Friction μ

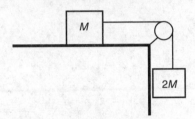

11. Coefficient of Friction μ

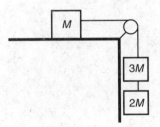

12. Frictionless

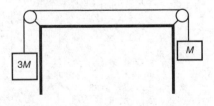

13. Frictionless

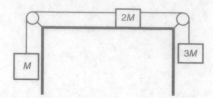

14. Coefficient of Friction μ

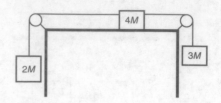

The Answers

1. $a = 10$ m/s^2

2. $a = 3.3$ m/s^2
   $T = 3.3$ N

See step-by-step solution below.

3. $a = 1.7$ m/s^2
   $T_1 = 1.7$ N
   $T_2 = 5.1$ N

4. $a = 1.3$ m/s^2
   $T = 3.3$ N

5. $a = 3.3$ m/s^2
   $T = 13$ N

See step-by-step solution below.

6. $a = 7.1$ m/s^2
   $T_1 = 17$ N
   $T_2 = 11$ N

7. $a = 3.3$ m/s^2
   $T = 6.6$ N

8. $a = 6.7$ m/s^2
   $T_1 = 13$ N
   $T_2 = 10$ N

9. $a = 1.7$ m/s^2
   $T_1 = 5.1$ N
   $T_2 = 8.3$ N

10. $a = 6.0$ m/s^2
   $T = 8.0$ N

11. $a = 8.0$ m/s^2
   $T_1 = 10$ N
   $T_2 = 4.0$ N

12. $a = 5.0$ m/s^2
   $T = 15$ N

13. $a = 3.3$ m/s^2
   $T_1 = 13$ N
   $T_2 = 20$ N

14. $a = 0.22$ m/s^2
   $T_1 = 20$ N
   $T_2 = 29$ N

Step-by-Step Solution to Problem 2
Step 1: Free-body diagrams:

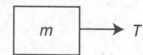   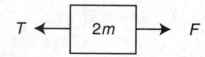

No components are necessary, so on to the next step.
Step 2: Write Newton's second law for each block, calling the rightward direction positive:

$$T - 0 = ma$$
$$F - T = (2m)a$$

Step 3: Solve algebraically. It's easiest to add these equations together, because the tensions cancel:

$$F = (3m)a, \text{ so } a = F/3m = (10 \text{ N})/3(1 \text{ kg}) = 3.3 \text{ m/s}^2.$$

To get the tension, just plug back into $T - 0 = ma$ to find $T = F/3 = 3.3$ N.

Step-by-Step Solution to Problem 5

Step 1: Free-body diagrams:

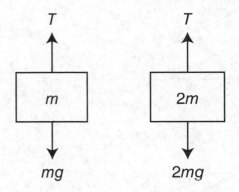

No components are necessary, so on to the next step.

Step 2: Write Newton's second law for each block, calling clockwise rotation of the pulley positive:

$$(2m)g - T = (2m)a$$
$$T - mg = ma$$

Step 3: Solve algebraically. It's easiest to add these equations together, because the tensions cancel:

$$mg = (3m)a, \text{ so } a = g/3 = 3.3 \text{ m/s}^2$$

To get the tension, just plug back into $T - mg = ma$: $T = m(a + g) = (4/3)mg = 13$ N.

Inclined Planes

How to Do It

Use the following steps to solve these kinds of problems: (1) Draw a free-body diagram for the object (the normal force is perpendicular to the plane; the friction force acts along the plane, opposite the velocity); (2) break vectors into components, where the parallel component of weight is $mg(\sin \theta)$; (3) write Newton's second law for parallel and perpendicular components; and (4) solve the equations for whatever the problem asks for.

Don't forget, the normal force is *not* equal to mg when a block is on an incline!

The Drill

Directions: For each of the following situations, determine:

(a) the acceleration of the block down the plane
(b) the time for the block to slide to the bottom of the plane

In each case, assume a frictionless plane unless otherwise stated; assume the block is released from rest unless otherwise stated.

1.

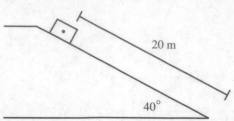

2.

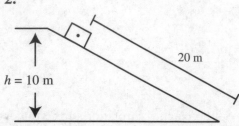

3.

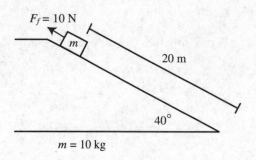

$m = 10$ kg

4.

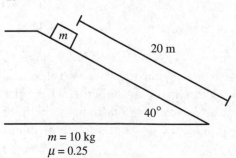

$m = 10$ kg
$\mu = 0.25$

5.

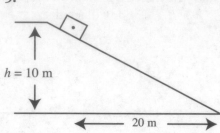

6.

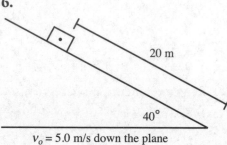

$v_o = 5.0$ m/s down the plane

7.

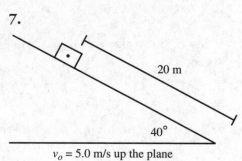

$v_o = 5.0$ m/s up the plane

8.

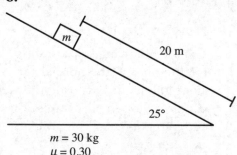

$m = 30$ kg
$\mu = 0.30$
$v_o = 3.0$ m/s up the plane

Careful—this one's tricky.

The Answers

1. $a = 6.3$ m/s^2, down the plane.
  $t = 2.5$ s
See step-by-step solution below.

2. $a = 4.9$ m/s^2, down the plane.
  $t = 2.9$ s

3. $a = 5.2$ m/s^2, down the plane.
  $t = 2.8$ s

4. $a = 4.4$ m/s^2, down the plane.
  $t = 3.0$ s

5. Here the angle of the plane is 27° by trigonometry, and the distance along the plane is 22 m.
  $a = 4.4$ m/s^2, down the plane.
  $t = 3.2$ s

6. $a = 6.3$ m/s^2, down the plane.
  $t = 1.8$ s

7. $a = 6.3$ m/s^2, down the plane.
  $t = 3.5$ s

8. This one is complicated. Since the direction of the friction force changes depending on whether the block is sliding up or down the plane, the block's acceleration is *not* constant throughout the whole problem. So, unlike problem 7, this one can't be solved in a single step. Instead, in order to use kinematics equations, you must break this problem up into two parts: up the plane and down the plane. During each of these individual parts, the acceleration is constant, so the kinematics equations are valid.

- up the plane:
  $a = 6.8$ m/s^2, down the plane.
  $t = 0.4$ s before the block turns around to come down the plane.

- down the plane:
  $a = 1.5$ m/s^2, down the plane.
  $t = 5.2$ s to reach bottom.

So, a total of $t = 5.6$ s for the block to go up and back down.

Step-by-Step Solution to Problem 1
Step 1: Free-body diagram:

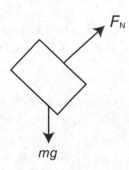

Step 2: Break vectors into components. Because we have an incline, we use inclined axes, one parallel and one perpendicular to the incline:

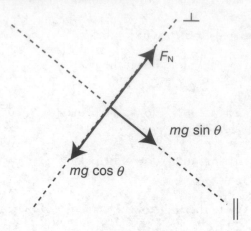

Step 3: Write Newton's second law for each axis. The acceleration is entirely directed parallel to the plane, so perpendicular acceleration can be written as zero:

$$mg \sin \theta - 0 = ma$$
$$F_N - mg \cos \theta = 0$$

Step 4: Solve algebraically for *a*. This can be done without reference to the second equation. (In problems with friction, use $F_f = \mu F_N$ to relate the two equations.)

$$a = g \sin \theta = 6.3 \text{ m/s}^2$$

To find the time, plug into a kinematics chart:

$$v_o = 0$$
$$v_f = \text{unknown}$$
$$\Delta x = 20 \text{ m}$$
$$a = 6.3 \text{ m/s}^2$$
$$t = ???$$

Solve for *t* using the second star equation for kinematics (**): $\Delta x = v_o t + \frac{1}{2} at^2$, where v_o is zero:

$$t = \sqrt{\frac{2\Delta x}{a}} = \sqrt{\frac{2(20 \text{ m})}{6.3 \text{ m/s}^2}} = 2.5 \text{ s}$$

Motion Graphs

How to Do It

For a position–time graph, the slope is the velocity. For a velocity–time graph, the slope is the acceleration, and the area under the graph is the displacement.

The Drill

Use the graph to determine something about the object's speed. Then play "Physics *Taboo*": suggest what object might reasonably perform this motion and explain in words how the object moves. Use everyday language. In your explanation, you may *not* use any words from the list below:

velocity
acceleration
positive
negative
increase
decrease
it
object
constant

(1)

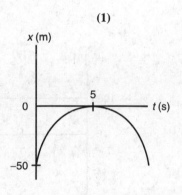

(2)

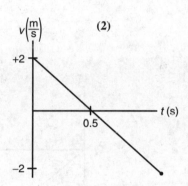

(3)

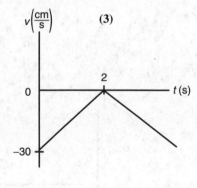

(4)

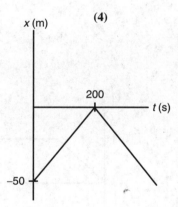

(5)

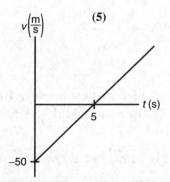

(6)

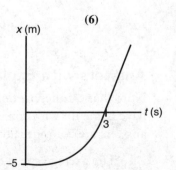

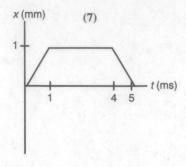

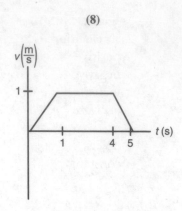

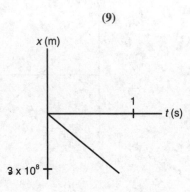

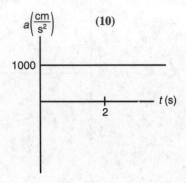

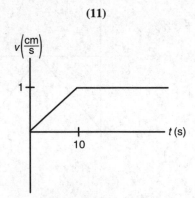

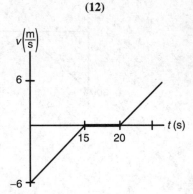

Answers with Explanations

Note that our descriptions of the moving objects reflect our own imaginations. You might have come up with some very different descriptions, and that's fine . . . provided that your answers are conceptually the same as ours.

1. The average speed over the first 5 s is 10 m/s, or about 22 mph. So:

Someone rolls a bowling ball along a smooth road. When the graph starts, the bowling ball is moving along pretty fast, but the ball encounters a long hill. So, the ball slows down, coming to rest after 5 s. Then, the ball comes back down the hill, speeding up the whole way.

2. This motion only lasts 1 s, and the maximum speed involved is about 5 mph. So:

A biker has been cruising up a hill. When the graph starts, the biker is barely moving at jogging speed. Within half a second, and after traveling only half a meter up the hill, the bike turns around, speeding up as it goes back down the hill.

3. The maximum speed of this thing is 30 cm/s, or about a foot per second. So:

A toy racecar is moving slowly along its track. The track goes up a short hill that's about a foot long. After 2 s, the car has just barely reached the top of the hill, and is perched there momentarily; then, the car crests the hill and speeds up as it goes down the other side.

4. The steady speed over 200 s (a bit over 3 minutes) is 0.25 m/s, or 25 cm/s, or about a foot per second.

A cockroach crawls steadily along the school's running track, searching for food. The cockroach starts near the 50 yard line of the football field; around three minutes later, the cockroach reaches the goal line and, having found nothing of interest, turns around and crawls at the same speed back toward his starting point.

5. The maximum speed here is 50 m/s, or over a hundred mph, changing speed dramatically in only 5 or 10 s. So:

A small airplane is coming in for a landing. Upon touching the ground, the pilot puts the engines in reverse, slowing the plane. But wait! The engine throttle is stuck! So, although the plane comes to rest in 5 s, the engines are still on . . . the plane starts speeding up backwards! Oops . . .

6. This thing covers 5 meters in 3 seconds, speeding up the whole time.

An 8-year-old gets on his dad's bike. The boy is not really strong enough to work the pedals easily, so he starts off with difficulty. But, after a few seconds he's managed to speed the bike up to a reasonable clip.

7. Though this thing moves quickly—while moving, the speed is 1 m/s—the total distance covered is 1 mm forward, and 1 mm back; the whole process takes 5 ms, which is less than the minimum time interval indicated by a typical stopwatch. So we'll have to be a bit creative:

In the Discworld novels by Terry Pratchett, wizards have developed a computer in which living ants in tubes, rather than electrons in wires and transistors, carry information. (Electricity has not been harnessed on the Discworld.) In performing a calculation, one of these ants moves forward a distance of 1 mm; stays in place for 3 ms; and returns to the original position. If this ant's motion represents two typical "operations" performed by the computer, then this computer has an approximate processing speed of 400 Hz times the total number of ants inside.

8. Though this graph *looks* like problem 7, this one is a velocity–time graph, and so indicates completely different motion.

A small child pretends he is a bulldozer. Making a "brm-brm-brm" noise with his lips, he speeds up from rest to a slow walk. He walks for three more seconds, then slows back down to rest. He moved forward the entire time, traveling a total distance (found from the area under the graph) of 4 m.

9. This stuff moves 300 million meters in 1 s at a constant speed. There's only one possibility here: electromagnetic waves in a vacuum.

   Light (or electromagnetic radiation of any frequency) is emitted from the surface of the moon. In 1 s, the light has covered about half the distance to Earth.

10. Be careful about axis labels: this is an *acceleration*–time graph. Something is accelerating at 1,000 cm/s² for a few seconds. 1,000 cm/s² = 10 m/s², about Earth's gravitational acceleration. Using kinematics, we calculate that if we drop something from rest near Earth, after 4 s the thing has dropped 80 m.

    One way to simulate the effects of zero gravity is to drop an experiment from the top of a high tower. Then, because everything that was dropped is speeding up at the same rate, the effect is just as if the experiment were done in the Space Shuttle—at least until everything hits the ground. In this case, an experiment is dropped from a 250-ft tower, hitting the ground with a speed close to 90 mph.

11. 1 cm/s is ridiculously slow. Let's use the world of slimy animals:

    A snail wakes up from his nap and decides to find some food. He speeds himself up from rest to his top speed in 10 s. During this time, he's covered 5 cm, or about the length of your pinkie finger. He continues to slide along at a steady 1 cm/s, which means that a minute later he's gone no farther than a couple of feet. Let's hope that food is close.

12. This one looks a bit like those up-and-down-a-hill graphs, but with an important difference—this time the thing stops not just for an instant, but for five whole seconds, before continuing back toward the starting point.

    A bicyclist coasts to the top of a shallow hill, slowing down from cruising speed (~15 mph) to rest in 15 s. At the top, she pauses briefly to turn her bike around; then, she releases the brake and speeds up as she goes back down the hill.

Simple Circuits

How to Do It

Think "series" and "parallel." The current through series resistors is the same, and the voltage across series resistors adds to the total voltage. The current through parallel resistors adds to the total current, and the voltage across parallel resistors is the same.

The Drill

For each circuit drawn below, find the current through and voltage across each resistor.
   Note: Assume each resistance and voltage value is precise to two significant figures.

1.

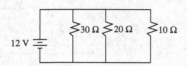

5.

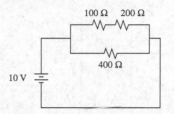

2.

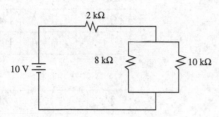

6.

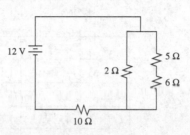

3.

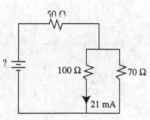

7.

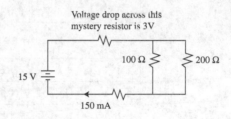

4.

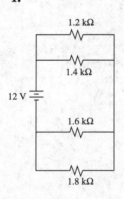

The Answers

1.

V	I	R
12 V	0.40 A	30 Ω
12 V	0.60 A	20 Ω
12 V	1.2 A	10 Ω
12 V	2.2 A	5.5 Ω

2.

V	I	R
3.2 V	1.6 mA	2 kΩ
6.8 V	0.9 mA	8 kΩ
6.8 V	0.7 mA	10 kΩ
10 V	1.6 mA	6.4 kΩ

(Remember, a kΩ is 1,000 Ω, and a mA is 10^{-3} A.)

See step-by-step solution following.

3.

V	I	R
2.5 V	0.051 A	50 Ω
2.1 V	0.021 A	100 Ω
2.1 V	0.030 A	70 Ω
4.6 V	0.051 A	91 Ω

4.

V	I	R
5.2 V	4.3 mA	1.2 kΩ
5.2 V	3.7 mA	1.4 kΩ
6.8 V	4.2 mA	1.6 kΩ
6.8 V	3.8 mA	1.8 kΩ
12 V	8.0 mA	1.5 kΩ

5.

V	I	R
3.4 V	0.034 A	100 Ω
6.8 V	0.034 A	200 Ω
10 V	0.025 A	400 Ω
10 V	0.059 A	170 Ω

6.

V	I	R
1.8 V	0.90 A	2.0 Ω
0.7 V	0.13 A	5.0 Ω
0.8 V	0.13 A	6.0 Ω
10.3 V	1.03 A	10.0 Ω
12.0 V	1.03 A	11.7 Ω

7.

V	I	R
3 V	0.15 A	20 Ω
10 V	0.10 A	100 Ω
10 V	0.05 A	200 Ω
2 V	0.15 A	13 Ω
15 V	0.15 A	100 Ω

Step-by-Step Solution to Problem 2

Start by simplifying the combinations of resistors. The 8 kΩ and 10 kΩ resistors are in parallel. Their equivalent resistance is given by

$$\frac{1}{R_{eq}} = \frac{1}{8\,k\Omega} + \frac{1}{10\,k\Omega}$$

which gives $R_{eq} = 4.4$ kΩ.

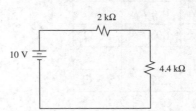

Next, simplify these series resistors to their equivalent resistance of 6.4 kΩ.

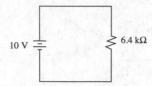

6.4 kΩ (i.e., 6,400 Ω) is the total resistance of the entire circuit. Because we know the total voltage of the entire circuit to be 10 V, we can use Ohm's law to get the total current

$$I_{total} = \frac{V_{total}}{R_{total}} = \frac{10\text{ V}}{6,400\ \Omega} = 0.0016\text{ A}$$

(more commonly written as 1.6 mA).

Now look at the previous diagram. The same current of 1.6 mA must go out of the battery, into the 2 kW resistor, and into the 4.4 kW resistor. The voltage across each resistor can thus be determined by V = (1.6 mA)R for each resistor, giving 3.2 V across the 2 kΩ resistor and 6.8 V across the 4.4 kΩ resistor.

The 2 kΩ resistor is on the chart. However, the 4.4 kΩ resistor is the equivalent of two parallel resistors. Because voltage is the same for resistors in parallel, there are 6.8 V across *each* of the two parallel resistors in the original diagram. Fill that in the chart, and use Ohm's law to find the current through each:

$$I_{8k} = 6.8\text{ V}/8,000\ \Omega = 0.9\text{ mA}$$
$$I_{10k} = 6.8\text{ V}/10,000\ \Omega = 0.7\text{ mA}$$

STEP 5

Build Your Test-Taking Confidence

AP PHYSICS 1 Practice Exam 1: Section I (Multiple-Choice)

AP PHYSICS 1 Practice Exam 1: Section II (Free-Response)

SOLUTIONS: AP Physics I Practice Exam 1, Section I (Multiple-Choice)

SOLUTIONS: AP Physics I Practice Exam 1, Section II (Free-Response)

AP PHYSICS 1 Practice Exam 2: Section I (Multiple-Choice)

AP PHYSICS 1 Practice Exam 2: Section II (Free-Response)

SOLUTIONS: AP Physics I Practice Exam 2, Section I (Multiple-Choice)

SOLUTIONS: AP Physics I Practice Exam 2, Section II (Free-Response)

Scoring the Practice Exams

Practice Exam 1

ANSWER SHEET FOR SECTION I

1 (A) (B) (C) (D)
2 (A) (B) (C) (D)
3 (A) (B) (C) (D)
4 (A) (B) (C) (D)
5 (A) (B) (C) (D)
6 (A) (B) (C) (D)
7 (A) (B) (C) (D)
8 (A) (B) (C) (D)
9 (A) (B) (C) (D)
10 (A) (B) (C) (D)
11 (A) (B) (C) (D)
12 (A) (B) (C) (D)
13 (A) (B) (C) (D)
14 (A) (B) (C) (D)
15 (A) (B) (C) (D)
16 (A) (B) (C) (D)
17 (A) (B) (C) (D)

18 (A) (B) (C) (D)
19 (A) (B) (C) (D)
20 (A) (B) (C) (D)
21 (A) (B) (C) (D)
22 (A) (B) (C) (D)
23 (A) (B) (C) (D)
24 (A) (B) (C) (D)
25 (A) (B) (C) (D)
26 (A) (B) (C) (D)
27 (A) (B) (C) (D)
28 (A) (B) (C) (D)
29 (A) (B) (C) (D)
30 (A) (B) (C) (D)
31 (A) (B) (C) (D)
32 (A) (B) (C) (D)
33 (A) (B) (C) (D)
34 (A) (B) (C) (D)

35 (A) (B) (C) (D)
36 (A) (B) (C) (D)
37 (A) (B) (C) (D)
38 (A) (B) (C) (D)
39 (A) (B) (C) (D)
40 (A) (B) (C) (D)
41 (A) (B) (C) (D)
42 (A) (B) (C) (D)
43 (A) (B) (C) (D)
44 (A) (B) (C) (D)
45 (A) (B) (C) (D)
46 (A) (B) (C) (D)
47 (A) (B) (C) (D)
48 (A) (B) (C) (D)
49 (A) (B) (C) (D)
50 (A) (B) (C) (D)

AP Physics 1 Practice Exam 1: Section I (Multiple-Choice)

Directions: The multiple-choice section consists of 50 questions to be answered in **90 minutes**. You may write scratch work in the test booklet itself, but only the answers on the answer sheet will be scored. You may use a calculator, the equation sheet, and the table of information.

Questions 1–45: Single-Choice Items

Directions: Choose the single best answer from the four choices provided and grid the answer with a pencil on the answer sheet.

1. A circuit consists of a battery and a light bulb. At first, the circuit is disconnected. Then, the circuit is connected, and the light bulb lights. After the light bulb has been lit for a few moments, how has the net charge residing on the circuit elements changed?

   (A) The net charge has become more positive.
   (B) The net charge has become more negative.
   (C) The net charge has not changed.
   (D) Whether the net charge becomes more positive or more negative depends on the initial net charge residing on the circuit elements before the bulb was lit.

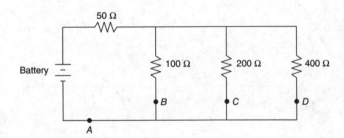

2. At which position in the above circuit will the charge passing that position in one second be largest?

   (A) A
   (B) B
   (C) C
   (D) D

3. Spring scales are used to measure the net force applied to an object; a sonic motion detector is used to measure the object's resulting acceleration. A graph is constructed with the net force on the vertical axis and the acceleration on the horizontal axis. Which of the following quantities is directly measured using the slope of this graph?

   (A) Gravitational mass
   (B) Weight
   (C) Velocity
   (D) Inertial mass

Questions 4 and 5 refer to the information below:

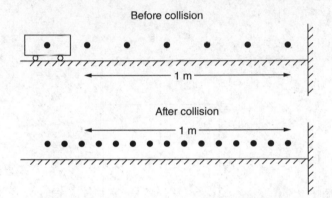

In the laboratory, a 0.5-kg cart collides with a fixed wall, as shown in the preceding diagram. The collision is recorded with a video camera that takes 20 frames per second. A student analyzes the video, placing a dot at the center of mass of the cart in each frame. The analysis is shown above.

4. About how fast was the cart moving before the collision?

   (A) 0.25 m/s
   (B) 4.0 m/s
   (C) 0.20 m/s
   (D) 5.0 m/s

5. Which of the following best estimates the change in the cart's momentum during the collision?

   (A) 27 N·s
   (B) 13 N·s
   (C) 1.3 N·s
   (D) 2.7 N·s

GO ON TO THE NEXT PAGE

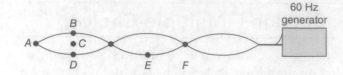

6. In the laboratory, a 60-Hz generator is connected to a string that is fixed at both ends. A standing wave is produced, as shown in the preceding figure. In order to measure the wavelength of this wave, a student should use a meterstick to measure from positions

(A) B to C
(B) B to D
(C) D to E
(D) A to F

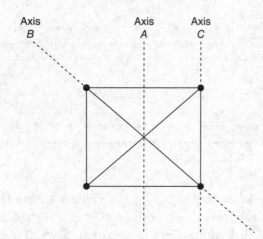

7. Four identical lead balls with large mass are connected by rigid but very light rods in the square configuration shown in the preceding figure. The balls are rotated about the three labeled axes. Which of the following correctly ranks the rotational inertia I of the balls about each axis?

(A) $I_B > I_A = I_C$
(B) $I_A > I_C = I_B$
(C) $I_C > I_A > I_B$
(D) $I_C > I_A = I_B$

8. In the laboratory, a cart experiences a single horizontal force as it moves horizontally in a straight line. Of the following data collected about this experiment, which is sufficient to determine the work done on the cart by the horizontal force?

(A) The magnitude of the force, the cart's initial speed, and the cart's final speed
(B) The mass of the cart and the distance the cart moved
(C) The mass of the cart, the cart's initial speed, and the cart's final speed
(D) The mass of the cart and the magnitude of the force

9. A wave pulse on a string is shown above. Which pulse, when superimposed with the one above, will produce complete destructive interference?

(A)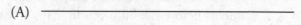

(B)

(C)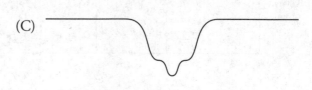

(D)

GO ON TO THE NEXT PAGE

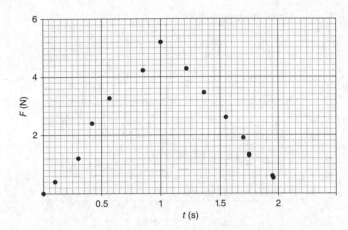

10. In the laboratory, a 3-kg cart experiences a varying net force. This net force is measured as a function of time, and the data collected are displayed in the graph above. What is the change in the cart's momentum during the interval $t = 0$ to $t = 2$ s?

(A) 5 N·s
(B) 10 N·s
(C) 15 N·s
(D) 30 N·s

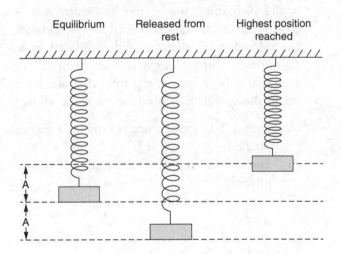

11. A block is attached to a vertical spring. The block is pulled down a distance A from equilibrium, as shown above, and released from rest. The block moves upward; the highest position above equilibrium reached by the mass is less than A, as shown. When the mass returns downward, how far below the equilibrium position will it reach?

(A) Greater than the distance A below equilibrium
(B) Less than the distance A below equilibrium
(C) Equal to the distance A below equilibrium
(D) No distance—the block will fall only to the equilibrium position.

Questions 12 and 13 refer to the following information:

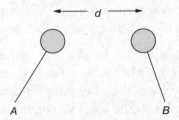

Two charged Styrofoam balls are brought a distance d from each other, as shown. The force on Ball B is 2 μN to the right. When the distance between the balls is changed, the force on Ball B is 8 μN to the right.

12. Which of the following can indicate the sign of the charges of balls A and B?

	Ball A	Ball B
(A)	positive	negative
(B)	neutral	positive
(C)	negative	negative
(D)	positive	neutral

13. When the force on Ball B is 8 μN, what is the distance between the centers of the two balls?

(A) $d/4$
(B) $d/2$
(C) $d/16$
(D) $d/\sqrt{2}$

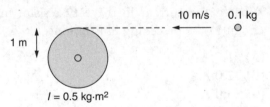

14. A disk of radius 1 m and rotational inertia $I = 0.5$ kg·m² is free to rotate, but initially at rest. A blob of putty with mass 0.1 kg is traveling toward the disk with a speed of 10 m/s, as shown in the preceding figure. The putty collides with the outermost portion of the disk and sticks to the disk. What is the angular momentum of the combined disk-putty system after the collision?

(A) 5 kg·m²/s
(B) 1 kg·m²/s
(C) 0.5 kg·m²/s
(D) 0 kg·m²/s

GO ON TO THE NEXT PAGE

15. A 1-kg object is released from rest at the top of a rough-surfaced incline. The object slides without rotating to the bottom of the incline. The object's kinetic energy at the bottom must be

(A) Equal to the block's gravitational potential energy when it was released, because total mechanical energy must be conserved.

(B) Equal to the block's gravitational potential energy when it was released, because the gain in kinetic energy compensates for the mechanical energy lost to thermal energy on the rough incline.

(C) Less than the block's gravitational potential energy when it was released, because the gravitational potential energy was converted both to thermal energy and to kinetic energy.

(D) Less than the block's gravitational potential energy when it was released, because the work done by the friction force must be greater than the block's gain in kinetic energy.

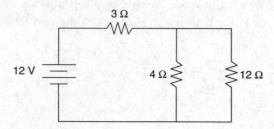

16. What is the current in the 4 Ω resistor in the circuit in the preceding illustration?

(A) 1.5 A
(B) 2.0 A
(C) 3.0 A
(D) 6.0 A

Oscilloscope Trace 1

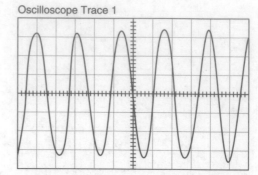

Oscilloscope Trace 2

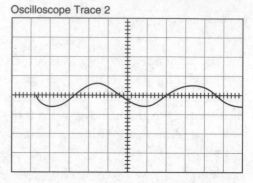

17. Two sounds are played in the laboratory. A microphone is connected to an oscilloscope, which displays the traces shown above for each sound. On these traces, the horizontal axis is time; the vertical axis is related to the distance the microphone's diaphragm is displaced from its resting position. The scales are identical for each diagram. Which of the following is correct about the sounds that produce the traces above?

(A) Sound 1 is louder, and sound 2 is higher pitched.

(B) Sound 2 is louder, and sound 2 is higher pitched.

(C) Sound 1 is louder, and sound 1 is higher pitched.

(D) Sound 2 is louder, and sound 1 is higher pitched.

18. The radius of Mars is about half that of Earth; the mass of Mars is about one-tenth that of Earth. Which of the following is closest to the gravitational field at the surface of Mars?

(A) 10 N/kg
(B) 4 N/kg
(C) 2 N/kg
(D) 0.5 N/kg

GO ON TO THE NEXT PAGE

Before

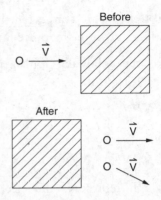

After

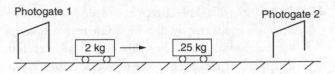

19. In an experiment, a marble rolls to the right at speed v, as shown in the top diagram. The marble rolls under a canopy, where it is heard to collide with marbles that were not initially moving. Such a collision is known to be elastic. After the collision, two equal-mass marbles are observed leaving the canopy with velocity vectors directed as shown. Which of the following statements justifies why the experimenter believes that a third marble was involved in the collision under the canopy?

(A) Before collision, the only marble momentum was directed to the right. After the collision, the combined momentum of the two visible marbles is still to the right. Another marble must have a leftward momentum component to conserve momentum.

(B) Before collision, the only marble momentum was directed to the right. After the collision, the combined momentum of the two visible marbles has a downward component; another marble must have an upward momentum component to conserve momentum.

(C) Before collision, the only marble kinetic energy was directed to the right. After the collision, the combined kinetic energy of the two visible marbles is still to the right. Another marble must have a leftward kinetic energy component to conserve kinetic energy.

(D) Before collision, the only marble kinetic energy was directed to the right. After the collision, the combined kinetic energy of the two visible marbles has a downward component; another marble must have an upward kinetic energy component to conserve kinetic energy.

20. In the laboratory, two carts on a track collide in the arrangement shown in the preceding figure. Before the collision, the 2-kg cart travels through photogate 1, which measures its speed; the 0.25-kg cart is initially at rest. After the collision, the carts bounce off one another. Photogate 2 measures the speed of each cart as it passes.

A student is concerned about his experimental results. When he adds the momentum of both carts after collision, he gets a value greater than the momentum of the 2-kg cart before collision. Which of the following is a reasonable explanation for the discrepancy?

(A) The track might have been slanted such that the carts were moving downhill.

(B) Human error might have been involved in reading the photogates.

(C) Friction might not have been negligible.

(D) The collision might not have been elastic.

	Wheel Structure	Wheel Mass	Wheel Radius
Wagon A	solid disk, $I = \frac{1}{2}MR^2$	0.5 kg	0.1 m
Wagon B	solid disk, $I = \frac{1}{2}MR^2$	0.2 kg	0.2 m
Wagon C	hollow hoop, $I = MR^2$	0.2 kg	0.1 m

21. Three wagons each have the same total mass (including that of the wheels) and four wheels, but the wheels are differently styled. The structure, mass, and radius of each wagon's wheels are shown in the preceding chart. In order to accelerate each wagon from rest to a speed of 10 m/s, which wagon requires the greatest energy input?

(A) Wagon A
(B) Wagon B
(C) Wagon C
(D) All require the same energy input

GO ON TO THE NEXT PAGE

22. A swimmer is able to propel himself forward through the water by moving his arms. Which of the following correctly states the applicant and recipient of the force responsible for the swimmer's forward acceleration?

(A) The force of the surrounding water on the swimmer's arms
(B) The force of the swimmer's arms on the swimmer's torso
(C) The force of the swimmer's arms on the surrounding water
(D) The force of the swimmer's torso on the swimmer's arms

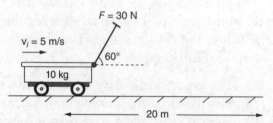

23. A 10-kg wagon moves horizontally at an initial speed of 5 m/s. A 30-N force is applied to the wagon by pulling the rigid handle, which is angled 60° above the horizontal. The wagon continues to move horizontally for another 20 m. A negligible amount of work is converted into thermal energy. By how much has the wagon's kinetic energy increased over the 20 m?

(A) 300 J
(B) 600 J
(C) 125 J
(D) 63 J

24. A moving 1.5-kg cart collides with and sticks to a 0.5-kg cart which was initially at rest. Immediately after the collision, the carts each have the same _____ as each other.

(A) Velocity
(B) Kinetic energy
(C) Mass
(D) Linear momentum

Questions 25 and 26 refer to the information below:

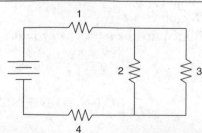

Four identical resistors are connected to a battery in the configuration shown in the preceding figure.

25. Which of the following ranks the current I through each resistor?

(A) $I_1 = I_4 > I_2 > I_3$
(B) $I_1 = I_4 > I_2 = I_3$
(C) $I_1 = I_2 = I_3 = I_4$
(D) $I_1 > I_2 = I_3 > I_4$

26. Which graph represents the electric potential with respect to the negative end of the battery as a function of the location on a loop of wire starting from the positive end of the battery, going through resistors 1, 2, and 4, and ending back on the negative end of the battery?

(A)
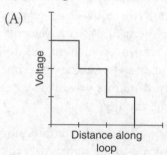

(B)
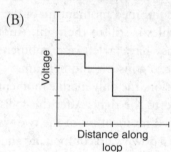

(C)
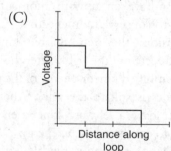

(D)
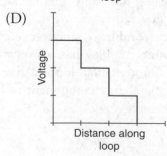

GO ON TO THE NEXT PAGE

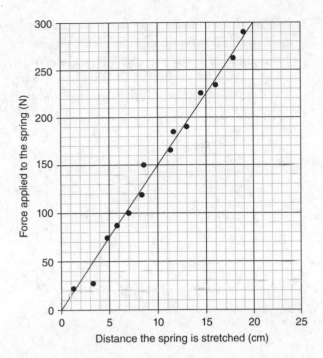

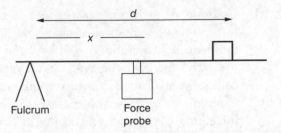

27. A force probe is used to stretch a spring by 20 cm. The graph of the force as a function of distance shown in the preceding figure is produced and used to determine the amount of work done in stretching the spring 20 cm. The experimenter reports the result as 3,000 N·cm. Which of the following is a reasonable estimate of the experimental uncertainty on this measurement?

(A) 3,000 ± 3 N·cm
(B) 3,000 ± 30 N·cm
(C) 3,000 ± 300 N·cm
(D) 3,000 ± 3,000 N·cm

28. A string of fixed tension and linear mass density is attached to a vibrating speaker. It is observed that a speaker frequency of 60 Hz does not produce standing waves in the string. Which explanation for this phenomenon is correct?

(A) The string length is not a multiple of half the wavelength of the wave.
(B) The wave speed on the string is fixed.
(C) 60 Hz is in the lowest range of audible sound.
(D) The wavelength of the wave produced by the speaker is equal to the speed of waves on the string divided by 60 Hz.

29. In the laboratory, a long platform of negligible mass is free to rotate on a fulcrum. A force probe is placed a fixed distance x from the fulcrum, supporting the platform. An object of fixed mass is placed a variable distance d from the fulcrum. For each position d, the force probe is read. It is desired to determine the mass of the object from a graph of data. Which of the following can determine the object's mass?

(A) Plot the reading in the force probe times x on the vertical axis; plot the gravitational field times d on the horizontal axis. The mass is the slope of the line.
(B) Plot the reading in the force probe on the vertical axis; plot the distance d on the horizontal axis. The mass is the area under the graph.
(C) Plot the reading in the force probe on the vertical axis; plot the distance d multiplied by the distance x on the horizontal axis. The mass is the y-intercept of the graph.
(D) Plot the reading in the force probe times d on the vertical axis; plot the distance x on the horizontal axis. The mass is the slope of the line divided by the gravitational field.

30. In Collision A, two carts collide and bounce off each other. In Collision B, a ball sticks to a rigid rod, which begins to rotate about the combined center of mass. Which of the following statements about quantities in each collision is correct?

(A) Collision A: each cart experiences the same force, time of collision, and change in kinetic energy. Collision B: the ball and the rod each experience the same torque, time of collision, and change in rotational kinetic energy.

GO ON TO THE NEXT PAGE

(B) Collision A: each cart experiences the same force, time of collision, and change in linear momentum. Collision B: the ball and the rod each experience the same torque, time of collision, and change in angular momentum.

(C) Collision A: each cart experiences the same force, time of collision, and change in kinetic energy. Collision B: the ball and the rod each experience the same torque, time of collision, and change in angular momentum.

(D) Collision A: each cart experiences the same force, time of collision, and change in velocity. Collision B: the ball and the rod each experience the same torque, time of collision, and change in angular velocity

31. It is known that a lab cart is moving east at 25 cm/s at time $t_1 = 0.10$ s, and then moving east at 15 cm/s at $t_2 = 0.20$ s. Is this enough information to determine the direction of the net force acting on the cart between t_1 and t_2?

(A) Yes, since we know the cart is slowing down, its momentum change is opposite the direction of movement, and the net force is in the direction of momentum change.

(B) No, because we don't know whether forces such as friction or air resistance might be acting on the cart.

(C) No, because we don't know the mass of the cart.

(D) Yes, since we know the cart keeps moving to the east, the net force must be in the direction of motion.

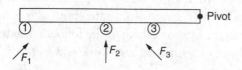

32. A rigid rod is pivoted at its right end. Three forces of identical magnitude but different directions are applied at the positions 1, 2, and 3 as shown. Which of the following correctly ranks the torques τ_1, τ_2, and τ_3 provided by the forces F_1, F_2, and F_3?

(A) $\tau_1 > \tau_2 > \tau_3$
(B) $\tau_3 > \tau_2 > \tau_1$
(C) $\tau_2 > \tau_1 > \tau_3$
(D) $\tau_2 > \tau_1 = \tau_3$

33. A block hanging vertically from a spring undergoes simple harmonic motion. Which of the following graphs could represent the acceleration a as a function of position x for this block, where $x = 0$ is the midpoint of the harmonic motion?

(A)
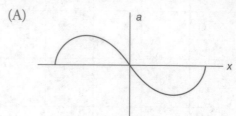

(B)
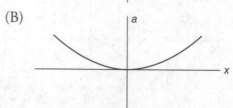

(C)
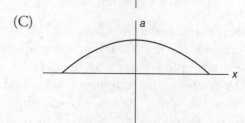

(D)
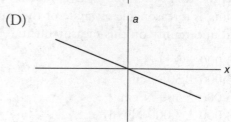

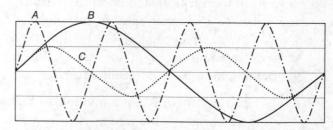

34. The preceding diagram represents a photograph of three transverse waves, each of which is moving to the right in the same material as the others. Which of the following ranks the waves by their amplitudes?

(A) A = B > C
(B) B > C > A
(C) A > C > B
(D) A = B = C

GO ON TO THE NEXT PAGE

35. The mass of the Earth is 5.97×10^{24} kg. The Moon, whose center is 3.84×10^8 m from the Earth's center, has mass 7.35×10^{22} kg. Which of the following is the best estimate of the gravitational force of the Earth on the Moon?

(A) 10^{39} N
(B) 10^{29} N
(C) 10^{19} N
(D) 10^9 N

36. A children's toy consists of a cart whose very light wheels are attached to a rubber band. This rubber band can wind and unwind around the axle supporting the wheels.

This toy is given a shove, after which the toy rolls across a flat surface and up a ramp. It is observed that the toy does not go a consistent distance up the ramp—in some trials it ends up higher than in other trials, even though the shove imparts the same kinetic energy to the cart each time. Which of the following is a reasonable explanation for this phenomenon?

(A) Depending on how the rubber band is initially wound, more or less potential energy can be transferred from the rubber band to the kinetic energy of the car's motion.
(B) The normal force on the cart's wheels will be different depending on how much the rubber band winds or unwinds.
(C) How much energy is transferred from kinetic energy to gravitational potential energy depends on the vertical height at which the cart ends up.
(D) Some of the cart's initial kinetic energy will be dissipated due to work done by friction.

37. A man stands on a platform scale in an elevator. The elevator moves upward, speeding up. What is the action-reaction force pair to the man's weight?

(A) The force of the elevator cable on the man
(B) The force of the man on the scale
(C) The force of the elevator cable on the elevator
(D) The force of the man on the Earth

38. The preceding diagram shows a speaker mounted on a cart that moves to the right at constant speed v. Wave fronts for the constant-frequency sound wave produced by the speaker are indicated schematically in the diagram. Which of the following could represent the wave fronts produced by the stationary speaker playing the same note?

(A)

(B)

(C)

(D)

39. A table supports a wooden block placed on the tabletop. Which fundamental force of nature is responsible for this interaction, and why?

(A) The electric force, because the protons in the nuclei of the top atomic layer of the table repel the nuclei in the bottom atomic layer of the wood.
(B) The gravitational force, because by $F = GMm/r^2$, the force of the table on the wood at that close range is sufficient to balance the force of the Earth on the wood.
(C) The electric force, because the outer electrons in the top atomic layer of the table repel the outer electrons in the bottom atomic layer of the wood.
(D) The strong nuclear force, because the protons in the nuclei of the top atomic layer of the table repel the nuclei in the bottom atomic layer of the wood.

GO ON TO THE NEXT PAGE

40. A solid sphere ($I = 0.06$ kg·m²) spins freely around an axis through its center at an angular speed of 20 rad/s. It is desired to bring the sphere to rest by applying a friction force of magnitude 2.0 N to the sphere's outer surface, a distance of 0.30 m from the sphere's center. How much time will it take the sphere to come to rest?

(A) 4 s
(B) 2 s
(C) 0.06 s
(D) 0.03 s

41. Which of the following force diagrams could represent the forces acting on a block that slides to the right while slowing down?

(A)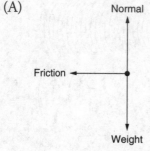

(B)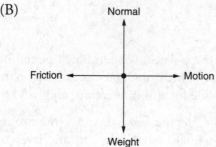

(C)

(D)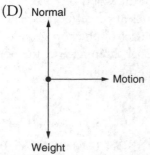

42. Standing waves are produced by a 100-Hz generator in a string of fixed length. The tension in the string is increased until a new set of standing waves is produced. Will the wavelength of the new standing waves be greater than or less than the wavelength of the original standing waves?

(A) Less, because the tension in the string varies directly with the wave speed, which varies inversely with the wavelength.

(B) Greater, because the tension in the string varies directly with the wave speed, which varies inversely with the wavelength.

(C) Greater, because the tension in the string varies directly with the wave speed, which varies directly with the wavelength.

(D) Less, because the tension in the string varies directly with the wave speed, which varies directly with the wavelength.

43. Two electrically charged balls are separated by a short distance, producing a force of 50 µN between them. Keeping the charge of each ball the same, the mass of one of the balls but not the other is doubled. What is the new electric force between the balls?

(A) 50 µN
(B) 100 µN
(C) 200 µN
(D) 400 µN

44. A block of mass m is attached to a spring of force constant k. The mass is stretched a distance A from equilibrium and released from rest. At a distance x from the equilibrium position, which of the following represents the kinetic energy of the block?

(A) $\frac{1}{2}kA^2 - \frac{1}{2}kx^2$
(B) $\frac{1}{2}m\,A\sqrt{k/m}\,^2$
(C) $\frac{1}{2}kA^2 + \frac{1}{2}kx^2$
(D) $\frac{1}{2}kx^2$

GO ON TO THE NEXT PAGE

45. A man stands with his hands to his sides on a frictionless platform that is rotating. Which of the following could change the angular momentum of the man-platform system?

(A) The man catches a baseball thrown to him by a friend.

(B) The man thrusts his arms out away from his body

(C) The man thrusts his arms out away from his body, and then quickly brings his arms back to his side again.

(D) The man jumps straight up in the air and lands back on the platform.

Questions 46–50: Multiple-Correct Items

Directions: Identify *exactly two* of the four answer choices as correct and grid the answers with a pencil on the answer sheet. No partial credit is awarded; both of the correct choices, and none of the incorrect choices, must be marked for credit.

46. The distance between the centers of two objects is d. Each object has identical mass m and identical charge $-q$. Choose all of the correct statements about the similarities and differences between the electric and gravitational force between the two objects. Choose two answers.

(A) Both the electric and the gravitational force depend inversely on the square of the distance d.

(B) Just as the gravitational force depends on the sum of the two masses m and m, the electric force depends on the sum of the two charges $-q$ and $-q$.

(C) For any measureable m and q in the laboratory, the electric force is many orders of magnitude larger than the gravitational force.

(D) Both the electric and gravitational forces are attractive.

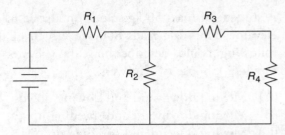

47. Which placement of voltmeters will allow for determination of the voltage across resistor R_2 in the circuit diagrammed in the preceding figure? Choose two answers.

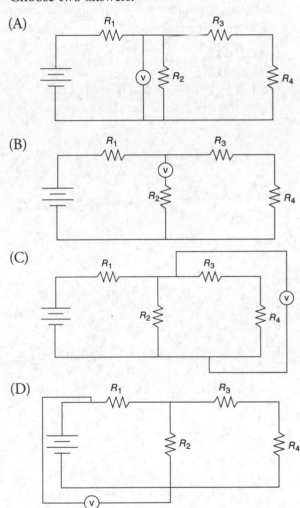

48. A student of mass 50 kg stands on a scale in an elevator. The scale reads 800 N. Which of the following could describe how the elevator is moving? Choose two answers.

(A) Moving downward and slowing down
(B) Moving downward and speeding up
(C) Moving upward and speeding up
(D) Moving upward and slowing down

49. A 1-m-long pipe is closed at one end. The speed of sound in the pipe is 300 m/s. Which of the following frequencies will resonate in the pipe? Choose two answers.

(A) 75 Hz
(B) 150 Hz
(C) 225 Hz
(D) 300 Hz

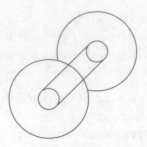

50. The device shown in the preceding figure consists of two wheels connected by a thick axle. A force can be applied to the axle by pulling a rope at several positions along the axle. Assuming the spool does not slip on the table, which of the pictured applied forces would cause rotation to the right of the device's wheels? Choose two answers.

Side view Side view Side view Side view
F_1 F_2 F_3 F_4

(A) F_1
(B) F_2
(C) F_3
(D) F_4

STOP. End of Physics 1 Practice Exam 1—Multiple-Choice Questions

AP Physics 1 Practice Exam 1: Section II (Free-Response)

Directions: The free-response section consists of five questions to be answered in 90 minutes. Budget approximately 20 to 25 minutes each for the first two longer questions; the next three shorter questions should take about 12 to 17 minutes each. Explain all solutions thoroughly, as partial credit is available. On the actual test, you will write the answers in the test booklet; for this practice exam, you will need to write your answers on a separate sheet of paper.

1. (12 points)

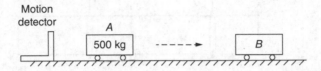

In Experiment 1, two carts collide on a negligible-friction track: Cart A with mass 500 g, and Cart B with unknown mass. Before the collision, Cart B is at rest. Adhesive is attached to the carts such that after the collision, the carts stick together. The speeds of Cart A before collision and after collision are measured using the sonic motion detector, as shown in the diagram.

(a) In one trial, the motion detector is turned on, Cart A is given a shove, the carts collide, and then the detector is turned off. The detector produces the velocity-time graph shown as follows. On the graph, indicate with a circle the portion of the graph that represents the collision occurring. Explain how you figured this out.

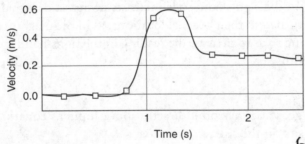

(i) Use the graph to estimate the speed of Cart A before the collision. 0.6
(ii) Use the graph to estimate the speed of Cart A after the collision. 0.27

(b) In numerous trials, the speeds of Cart A before and after the collision are measured. You are asked to construct a graph of this data whose slope can be used to calculate the mass of Cart B. P=mv
(i) What should you graph on each axis?
(ii) Explain in several sentences how you will use the slope of this graph to calculate the mass of Cart B. Be specific both about the calculations you will perform, and about why those calculations will produce the mass of Cart B.

(c) In Experiment 2, the adhesive is removed such that the carts bounce off of one another. The motion detector is again positioned to read the speed of Cart A before and after collision.
(i) Describe an experimental procedure by which the speed of Cart B after collision can be measured. You may use any equipment available in your physics laboratory, but you may *not* use a second sonic motion detector.

GO ON TO THE NEXT PAGE

(ii) The masses of Carts A and B are now both known; the speeds of both carts before and after collision have been measured. Explain how you could determine whether the collision in Experiment 2 was elastic. Be sure to describe specifically the calculations you would perform, as well as how you would use the results of those calculations to make the determination.

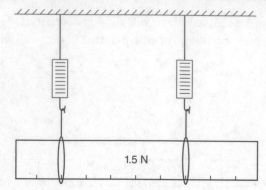

2. (12 points)

A uniform meterstick, which weighs 1.5 N, is supported by two spring scales. One scale is attached 20 cm from the left-hand edge; the other scale is attached 30 cm from the right-hand edge, as shown in the preceding diagram.

(a) Which scale indicates a greater force reading? Justify your answer qualitatively, with no equations or calculations.
(b) Calculate the reading in each scale.
(c) Now the right-hand scale is moved closer to the center of the meterstick but is still hanging to the right of center. Explain your answers to the following in words with reference to your calculations in (b).
 (i) Will the reading in the left-hand scale increase, decrease, or remain the same?
 (ii) Will the reading in the right-hand scale increase, decrease, or remain the same?
(d) Now the scales are returned to their original locations, as in the diagram. Where on the meterstick could a 0.2-N weight be hung so as to increase the reading in the right-hand spring scale by the largest possible amount? Justify your answer.

3. (7 points)

Space Probe A orbits in geostationary orbit directly above Jupiter's equator, 90,000 km above the surface. Identical Space Probe B sits on the surface of Jupiter.

(a) Which probe, if either, has a greater speed? Justify your answer
(b) Which probe, if either, has the greatest acceleration toward the center of Jupiter? Justify your answer.
(c) Consider the list of gravitational forces below:
 1. The force of Jupiter on Space Probe A
 2. The force of Jupiter on Space Probe B
 3. The force of Space Probe A on Jupiter
 4. The force of Space Probe B on Jupiter
 5. The force of Space Probe A on Space Probe B
 6. The force of Space Probe B on Space Probe A

 (i) Rank the magnitudes of the six gravitational forces listed above from greatest to least. If two or more quantities are the same, indicate so clearly in your ranking.

 Greatest _____ _____ _____ _____ _____ _____ *Least*

 (ii) Justify your ranking.

4. (7 points)

Two blocks, Block A of mass m and Block B of mass $2m$, are attached together by a spring. The blocks are free to move on a level, frictionless surface. The spring is compressed and then the blocks are released from rest.

    Consider two different systems. One system consists *only* of Block A; the other system consists of both blocks and the connecting spring. In a clear, coherent, paragraph-length response, explain whether kinetic energy, total mechanical energy, and/or linear momentum is conserved in each of the systems described.

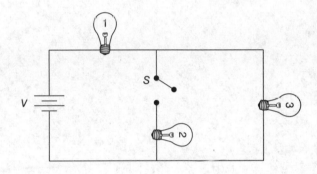

5. (7 points)

Three identical light bulbs are connected in the circuit shown above. The switch S is initially in the open position and then is closed at time t.

(a) Describe any changes that occur in the current through each bulb when the switch is closed. Justify your answers.

(b) Describe any changes that occur in the brightness of each bulb when the switch is closed. Justify your answers.

(c) When the switch is closed, does the power output of the battery increase, decrease, or remain the same? Justify your answer.

(d) When the switch is closed, the current in Bulb 1 changes. Explain why this change in current does not violate the law of conservation of charge.

Solutions: AP Physics 1 Practice Exam 1, Section I (Multiple-Choice)

Questions 1–45: Single-Choice Items

1. **C**—A circuit allows charge to flow and changes the electrical energy of the charges in the wires; however, by conservation of charge, the net charge of the circuit can't change unless some path is provided for charge to flow out of the circuit.

2. **A**—Kirchoff's junction rule, a statement of conservation of electric charge in a circuit, says that current entering a junction equals current leaving a junction. Current is charge flowing per second. The charge passing point A in one second must be equal to the sum of the charge passing B, C, and D in one second; A must have the greatest charge flow per second.

3. **D**—The relevant equation connecting force and mass is $F = ma$. The slope is the vertical axis divided by the horizontal axis, or $F/a = m$. So the slope is mass—but why inertial not gravitational mass? Inertia is defined as an object's resistance to acceleration. If acceleration is involved, you're talking inertial mass. Gravitational mass would involve the weight of an object in a gravitational field.

4. **B**—The dots divide the 1-meter distance into five parts. In the time between dots, the cart travels 1/5 of a meter, or 0.2 m. The time between dots is 1/20 of a second, or 0.05 s. At constant speed, the speed is given by distance/time: 0.20 m/0.05 s = 4 m/s.

5. **D**—Initially, the cart's mass is 0.5 kg and speed is 4 m/s, so the cart's momentum is $mv = 2$ N·s. In the collision, the cart loses that 2 N·s in order to stop briefly and then gains more momentum in order to speed up again. So the momentum change must be more than 2 N·s. How much more? After collision, the cart is moving slower than 4 m/s because the dots are closer together, so the cart's momentum is less than 2 N·s. The cart's momentum change is (2 N·s) + (something less than 2 N·s); the only possible answer is 2.7 N·s.

6. **D**—On a standing wave, the wavelength is measured from node-to-node-to-node (i.e., across two "humps").

7. **D**—The rotational inertia of a point mass is MR^2, where R is the distance from the mass to the axis of rotation. Pretend the side of the square is of length 2 m, and that each mass is 1 kg. For axis A, each mass has rotational inertia (1 kg) $(1 \text{ m})^2 = 1$ kg·m². With four masses total, that's **4** kg·m². For axis B, each mass is $\sqrt{2}$ m from the axis (the diagonal of the square is $2\sqrt{2}$ m, each mass is half a diagonal from the axis). Each mass has $(1 \text{ kg})(\sqrt{2}\text{m})^2 = 2$ kg·m². Two masses make a total of **4** kg·m². And for axis C, the masses are each 2 m from the axis, so they each have (1 kg) $\times (2 \text{ m})^2 = 4$ kg·m². With two masses, that's a total of **8** kg·m². So this would be ranked axis C, followed by equal axes A and B.

8. **C**—The work done by the force is force times distance; but that's not an option. The other option to find work is that work done by a nonconservative force is equal to the change in an object's potential and kinetic energy. There's no potential energy change, because the surface is horizontal. The kinetic energy $(\frac{1}{2}mv^2)$ change can be determined by knowing the mass and the speed change.

9. **B**—When two pulses superimpose, the amplitudes add algebraically. That means that a part of the wave pulse on the top of the rope is canceled by a part of the wave pulse on the bottom of the rope. Since the middle part of the wave pulse is closer to the rope than the two humps, the middle hump of Choice B will cancel this pulse—its middle part is also closer to the rope than its humps.

10. **A**—Change in momentum is also known as impulse and is equal to force times time interval. On this graph, the multiplication of the axes means to take the area under the graph. Each segment of the data looks like it represents a straight line, making a big triangle. The area of a triangle is ½(base)(height). That's ½(5 N)(2 s) = 5 N·s.

11. **B**—If friction and air resistance are negligible, a mass on a spring oscillates about the equilibrium position, reaching the same maximum distance above and below. In this case, since the

mass doesn't get all the way to position A at the top, mechanical energy was lost (to friction or air resistance or some nonconservative force). Thus, without some external energy input, the mass won't reach its maximum position at the bottom, either—at the bottom it will have no kinetic energy, so all the energy will be potential, and we've already established that some total mechanical energy was lost.

12. **C**—If Ball B is forced to the right by Ball A, then Ball A must be forced to the left by Ball B—that's Newton's third law. These balls repel—only like-signed charges repel.

13. **B**—Coulomb's law gives the force of one charge on another, $F = k\dfrac{Q_1 Q_2}{r^2}$. We haven't changed the charges Q; it's the force of A on B that's quadrupled. Since the force F has increased, the distance between charges r has decreased. Since the force has quadrupled, and since the r in the denominator is square, the distance has been cut in half $(1/2)^2 = 1/4$; and making the denominator four times smaller makes the whole fraction four times bigger.

14. **B**—In a collision, momentum—including angular momentum—is conserved. The question might as well be asking, "What is the angular momentum of the two objects before the collision?" And since the disk is at rest initially, the question is asking the even easier question, "What is the angular momentum of the putty before collision?" The axis of rotation is the center of the disk. The putty is a point mass; the angular momentum of a point mass is mvr with r the distance of closest approach to the axis. That's $(0.1 \text{ kg})(10 \text{ m/s})(1 \text{ m}) = 1 \text{ kg·m}^2/\text{s}$.

15. **C**—Mechanical energy is not conserved because of the work done by the nonconservative friction force provided by the rough-surfaced incline. The object starts with only gravitational potential energy, because it is higher than its lowest position and at rest. This gravitational energy is converted to thermal energy via the friction force, and to kinetic energy because the object speeds up.

16. **A**—First simplify the 4 Ω and 12 Ω parallel combination to a 3 Ω equivalent resistance. In series with the other 3 Ω resistance, that gives a total resistance for the circuit of 6 Ω. By Ohm's law used on the whole circuit, $12 \text{ V} = I(6 \text{ Ω})$. The total current in the circuit is thus 2 A. This current must split between the two parallel resistors. The only possible answer, then, is 1.5 Ω—the current in the 4 Ω resistor must be less than the total current.

17. **C**—The amplitude of a sound is related to the loudness. Sound 1 has higher amplitude, so it is louder. The pitch of a sound is related to the frequency. Since the horizontal axis is time, the peak-to-peak distance corresponds to a period, which is the inverse of the frequency. Sound 1 has a smaller period, so it has a larger frequency and a larger pitch.

18. **B**—The gravitational field at the surface of a planet is $g = G\dfrac{M}{r^2}$. The numerator for Mars is 1/10 that of Earth, reducing the gravitational field by 1/10. The denominator for Mars is $(1/2)^2 = 1/4$ that of Earth, increasing the gravitational field by a factor of 4 (because a smaller denominator means a bigger fraction). The overall gravitational field is multiplied by 4/10. On Earth, the gravitational field is 10 N/kg, so on Mars, $g = 4$ N/kg.

19. **B**—Choices C and D are wrong because kinetic energy doesn't have a direction. Choice A is wrong because momentum conservation does not require a leftward momentum component—since the initial momentum was all to the right, the final momentum should be to the right. It's the vertical momentum that's the problem. Since the vertical momentum was zero to start with, any vertical momentum after collision must cancel out.

20. **A**—Choice B is ridiculous—scientists should never refer generically to "human error." Significant friction should reduce, not increase, the speed (and thus the momentum) measured by Photogate 2. The elasticity of a collision refers to kinetic energy conservation, not momentum conservation—even inelastic collisions must conserve momentum. If the track is slanted downhill to the right, then the carts speed up; conservation of momentum won't be valid between Photogates 1 and 2 because the downhill component of the gravitational force is a force external to the two-cart system.

21. **B**—The energy input must be enough to change the translational kinetic energy of the cart *and* to change the rotational kinetic energy of the wheels. Since all carts have the same mass and change speed by the same amount, they all require the same energy input to change the translational KE. Whichever wheels have the largest rotational inertia will require the largest energy input to get to the same speed. Calculating, wagon B has the largest rotational inertia of 0.004 kg·m².

22. **A**—Choice C is not correct because if the *swimmer* is accelerating, the responsible force must act on the swimmer, not on something else. That force can act on any part of the swimmer's body. But a force provided by the swimmer himself on the swimmer himself won't accelerate him—that's like pulling yourself up by your own bootstraps. Choices B and D do not consider a force external to the swimmer. The answer is A: the Newton's third law force pair to the force of the swimmer's arms on the water.

23. **A**—The work-energy theorem says that the work done by a nonconservative force is equal to the change in potential energy plus the change in kinetic energy. Since the wagon is on a horizontal surface, the potential energy change is zero; the work done by the 30-N pulling force is the change in the wagon's KE. Work is force times parallel displacement, so we don't use 30 N in this formula, we use the component of the 30-N force parallel to the 20-m displacement. That's (30 N)(cos 60)(20 m) = 300 N.

24. **A**—Yes, momentum is conserved. That means the carts combined have the same momentum as they did in sum before the collision. And sure, if we were solving for the speed after collision, we'd combine the masses together for the calculation, but that doesn't mean that the carts both have the same mass—one cart is 1.5 kg, the other 0.5 kg. The problem asks what is the same for each cart in comparison to the other. Since the carts are stuck together, they must move together. They have the same velocity as one another.

25. **B**—Current can only travel through a wire. At the junction after Resistor 1, the current splits; the current comes back together after resistors 2 and 3. Thus, resistors 1 and 4 have the same current that's equal to the total coming from the battery. Since resistors 2 and 3 are identical, they split the current evenly.

26. **D**—Since resistors 1 and 4 take the same current and are identical, the voltage drops the same amount across them. Since Resistor 2 takes less current than the others but has the same resistance, by $V = IR$ the voltage drops less across Resistor 2 than the others. Choice A is wrong because all voltage drops are the same. Choices B and C have Resistor 2 dropping the voltage by more than at least one of the other resistors. Choice D includes the same voltage drop across resistors 1 and 4, but a smaller drop across 2.

27. **C**—The work done by the spring is the area under a force-distance graph because work = force times distance. Using the best-fit line drawn as the top of a triangle, the area is (1/2) × (300 N)(20 cm) = 3,000 N·cm.* Now, put your ruler along the data points. Try to draw another line that's still a reasonable best fit, but is a bit shallower. Where does that line intersect the 20-cm position? It intersects at a point probably not much below 280 N, maybe even 290 N. The smallest possible work done, given this data, would be area = (1/2)(280 N)(20 cm) = 2,800 N·cm, which is 200 N·cm short of the 3,000 N·cm original estimate. That's closest to Choice C. If you've done a lot of in-class lab work, you might have noticed that your data often look about as scattered as shown in the graph; and that anything you calculate is never much closer to a known value or to your classmates' calculations than 5 or 10 percent. Here, Choice B works out to an uncertainty of 1 percent; Choice D is 100 percent. So C is the reasonable choice.

28. **A**—Standing waves on a string only exist when the wave pattern produces a node at each end. A wavelength is measured node-to-node-to-node; so half a wavelength is node-to-node. If the string length isn't a multiple of this node-to-node distance, then some fraction of a wave pattern would be left on one end rather than a node. That can't happen on a fixed string, so no standing waves exist.

29. **A**—Since the platform itself is of negligible mass, only two torques act on the platform:

---

*The units are N·cm rather than joules because a joule is a newton times a meter, not a newton times a centimeter.

counterclockwise by the force of the force probe (F_P), and clockwise by the downward force of the object (mg). Torque is force times distance from a fulcrum, so set these torques equal: $F_P(x) = mg(d)$. Solving for the mass, we get $\dfrac{F_p x}{gd}$. Plot the numerator on the vertical, the denominator on the horizontal, and the slope is the mass m.

30. **B**—The problem says nothing about the collision being "elastic"; therefore, the change in kinetic energy of any sort does not have to be the same for each object. Velocity is never conserved in a collision—momentum is—so Choice D is ridiculous. Newton's third law demands that the force of one cart on another is the same, and so also the torque on each object about the center of mass.

31. **A**—Net force includes the contributions of all forces, including friction or air resistance; and net force is in the direction of *acceleration*, not of motion.

32. **A**—Torque is force times distance from a fulcrum; but that force must be perpendicular to the rod, so in this case the force used in the equation will be the vertical component, which includes a sin 45° term for F_1 and F_3. The sine of 45° is 0.7; call the length of the rod L, so F_2 is a distance $L/2$ from the pivot, and F_3 is about $L/4$ from the pivot. So $\tau_1 = 0.7FL$. $\tau_2 = 0.5FL$. $\tau_3 = (0.7 \cdot 0.25)FL$.

33. **D**—The net force of a spring is kx, and so changes linearly with distance (because the x is not squared or square rooted). Acceleration is related to the net force, so acceleration also changes linearly with distance. The force is always toward the equilibrium position: when the block is pulled down, it is forced (and thus accelerates) up; when the block is pushed up, it accelerates down. So a negative x gives a positive a, as in (D).

34. **A**—The amplitude is measured from the midpoint to the peak or trough of a wave. Waves A and B are each two "bars" above the midpoint, while C is only one bar above the midpoint.

35. **C**—Use the equation $F = G\dfrac{M_1 M_2}{d^2}$ but just use the power of 10 associated with the value—the

answer choices are so far separated that more precision would be useless. $F = \dfrac{(10^{-11})(10^{22})(10^{24})}{(10^8)^2}$. Add exponents in the numerator because everything is multiplied together: $F = \dfrac{(10^{35})}{10^{16}}$. Now subtract exponents for the division problem: $F = 10^{19}$ N.

36. **A**—Choice B is wrong—the normal force on the flat surface is equal to the cart's weight, regardless of the rubber band. Choice C is true but does not explain different heights in each trial—the problem said that the kinetic energy provided to the cart was the same every time. Choice D may or may not be true but is irrelevant in any case—even if kinetic energy is lost to work done by friction, neither the force of nor the coefficient of friction changes in different trials, so that can't explain different heights. Now, the rubber band, though, that can change things. If it's initially wound and able to unwind as the cart moves, it can transfer some of its elastic potential energy to kinetic energy of the cart. Or, if it's initially unwound, it will require some kinetic energy in order to wind up again and store elastic potential energy.

37. **D**—The man's weight is the force of the Earth on the man. The Newton's third law force pair is then the force of the man on the Earth.

38. **C**—The Doppler effect states that a moving speaker will cause a stationary observer to hear a higher frequency. This is because the wave fronts will be emitted closer together than if the speaker were stationary—the waves will seem "scrunched" together. Well, for the wave fronts to be "scrunched" in the diagram for the moving cart, they'd have to be farther apart in the diagram for the stationary cart.

39. **C**—Choice B is ridiculous because not only is gravity the weakest of the fundamental forces, it makes no sense that the gravitational force of the maybe 100-kg table is similar to the gravitational force of the 10^{24} kg Earth in the equation $F = G\dfrac{M_1 M_2}{d^2}$. Choices A and D are ridiculous because the protons are about 10^5 (100,000) times smaller than the diameter of an atom—any interactions between atoms have to involve the electrons at the outside of the atoms, not protons far away in the center.

40. B—This is a calculation using $\tau_{net} = I\alpha$. The net torque on the sphere is force times distance from the center, or $(2.0\ N)(0.30\ m) = 0.60\ m\cdot N$. Now the angular acceleration can be calculated: $(0.60\ m\cdot N) = (0.06\ kg\cdot m^2)(\alpha)$, so $\alpha = 10\ rad/s$ per second. Use the definition of angular acceleration: the sphere loses 10 rad/s of speed each second. It started with 20 rad/s of speed, so after 1 s it has 10 rad/s of speed; after 2 s it will have lost all its speed.

41. A—There's no such thing as the "force of motion," so motion does not belong on a force diagram. The block slows down, so the block has acceleration in the opposite direction of motion, or left. That means there must be a leftward force.

42. C—The speed of waves on a string is given by $v = \sqrt{\dfrac{T}{m/L}}$. All we really need to know is that the bigger the tension, the bigger the wave speed—we know that because tension is in the numerator. Then use $v = \lambda f$: wavelength is in the numerator, so wave speed varies directly with wavelength.

43. A—The electric force of one charge on another is given by $F = k\dfrac{Q_1 Q_2}{d^2}$. The mass is not in this equation; doubling the mass does nothing to the electric force.

44. A—Here use the work-energy theorem, $W_{NC} = \Delta PE + \Delta KE$. The change in kinetic energy is what we're looking for, since the block began at rest. The only force doing work is the spring force, and that's a conservative force, so $W_{NC} = 0$. What we want, then, is the change in the block's potential energy, or the initial and final potential energies subtracted from one another. The potential energy of a block on a spring is $\frac{1}{2}kx^2$. What values do we use for x here? The initial and final positions are used. At first the block is at position $x = A$. At the end, the block is at a distance called x from equilibrium.

45. A—Angular momentum can only change when a torque that's *external* to the man-platform system acts; that is, when a torque is applied by something that isn't the man or the platform. The force of the baseball on the man acting anywhere but precisely at the center of the man's rotational motion will provide a torque, and thus change the man-platform's angular momentum. The other choices all involve interactions only between the man and the platform.

Questions 46–50: Multiple-Correct Items (You must indicate both correct answers; no partial credit is awarded.)

46. A and **C**—A is right because of the equations $F = k\dfrac{Q_1 Q_2}{d^2}$ and $F = G\dfrac{M_1 M_2}{d^2}$. B is incorrect—the numerators involve the product, not the sum, of masses and charges. C is correct; you could memorize that fact, or you could try plugging in the powers of 10 for measurable masses and charges. Try charges as small as nanocoulombs, or 10^{-9} C, and milligrams, or 10^{-6} kg. Since k for the electric force equation is 10^9 N·m²/C², and G for the gravitational force equation is 10^{-11} N·m²/kg², the electric force will always come out much bigger. D is incorrect because while gravity always attracts, only opposite-signed charges attract.

47. A and **C**—This is essentially Kirchoff's loop rule, which boils down to "items in parallel take the same voltage." It's pretty easy to see that the voltmeter in (A) is in parallel with R_2, but in (C) it's harder. Note that the current must divide between R_2 and the voltmeter, and then the current comes immediately back together after going through the voltmeter and R_2. That's parallel. The combination of R_3 and R_4 are *also* in parallel with the voltmeter here, but that's not relevant to the problem. Choice (B) is wrong because the voltmeter is in series with R_2. Choice (D) is wrong because the voltmeter is in parallel with the battery, which is not in parallel with R_2, because R_1 is in the way.

48. A and **C**—A free-body diagram would include the student's 500-N (50 kg·g) weight downward, and the scale's 800-N force upward. (No, there's no "force of the elevator." The elevator isn't in contact with the student.) So the net force is upward; and the acceleration is in the direction of the net force, also upward. Upward acceleration could mean one of two things: speeding up and moving upward, or slowing down and moving downward.

49. **A** and **C**—The fundamental frequency of a standing wave on a pipe closed at one end is $\frac{v}{4L}$. Here, with $v = 300$ m/s and $L = 1$ m, the fundamental frequency is 75 Hz. On a closed pipe, odd multiples of the fundamental frequency will resonate. So 75 Hz, $3 \cdot 75$ Hz $= 225$ Hz, and $5 \cdot 75$ Hz $= 375$ Hz, with nothing in between.

50. **A** and **D**—The key is knowing where the "fulcrum," or the pivot for rotation, is. Here, that's the contact point between the surface and the wheel. Now look at the line of the force. In F_1, the line of force is pulling right above the pivot point, and that causes clockwise, or forward, rotation. For F_2, the line of force is pulling up to the right of the pivot, and that causes counterclockwise, or backward, rotation. F_3's line of force is going right through the pivot point, so it will cause no rotation at all. F_4 pulls up with a line of force to the left of the pivot. This produces clockwise, or forward, rotation.

Solutions: AP Physics 1 Practice Exam 1, Section II (Free-Response)

Obviously your solutions will not be word-for-word identical to what is written below. Award points for your answer as long as it contains the correct physics, and as long as it does *not* contain incorrect physics.

Question 1

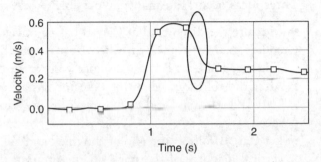

Part (a)

1 point for a correct circle on the diagram.

1 point for a correct explanation: When the carts collide, the moving cart must lose speed. The vertical axis of the velocity-time graph indicates speed. The circled portion of the graph is the only place where the vertical axis value drops rapidly, as the cart's speed must drop in the collision.

(i) **1 point** for answer: 0.60 m/s (or thereabouts—anywhere between, say, 0.55 m/s and 0.60 m/s is fine.)

(ii) **1 point** for answer: 0.25 m/s (or thereabouts—anything between, say, 0.22 m/s and 0.29 m/s is fine.)

Part (b)

(i) **1 point:** The easiest answer is to put the speed of Cart A before collision on the vertical axis; and to put the speed of Cart A after collision on the horizontal axis. There are other answers that will work.

(ii) **3 points:** Award one or two points of partial credit for correct but incomplete physics. For example, writing and using an expression for momentum conservation should earn a point, even if the rest of the explanation doesn't follow properly.

Conservation of momentum means the total momentum before the collision equals the total momentum after the collision. Before the collision, the total momentum is that of Cart A: $m_A v_A$. After the collision, the total momentum is $(m_A + m_B)v'$ where v' is the speed of Cart A (and, because they stick together, the speed of Cart B, too). Set these momentum expressions equal:

$$m_A v_A = (m_A + m_B)v'.$$

This equation can be solved for the *y*-axis variable divided by the *x*-axis variable:

$$\frac{v_A}{v'} = \frac{(m_A + m_B)}{m_A}.$$

So, to get the mass of Cart B, I'd determine the slope of the line on the graph, and set that equal to $\frac{(m_A + m_B)}{m_A}$. The mass of Cart A is given as 500 g, so I'd plug that in and solve for m_B.

Part (c)

(i) **2 points:** Award one point for a partially correct description, two points for a complete and correct description.

Three ideas occur, but many are possible:

- Measure the distance that Cart B has to travel to the end of the track. When the carts collide, start a stopwatch; when Cart B hits the end of the track, stop the stopwatch. The speed of Cart B is the distance you measured divided by the time on the stopwatch.
- After the collision, when the detector has already read the speed of Cart A but before Cart B reaches the end of the track, lift up Cart A. Now the detector can read Cart B's speed.
- Let Cart B roll off the end of the track and fall to the floor as a projectile. Measure the vertical height y of the track off the ground; the time t that the cart was in the air is given by $y = \frac{1}{2}gt^2$, where g is 10 m/s per second. Measure the horizontal distance from the track's edge to the spot where the cart landed. Then the speed of the cart is this horizontal distance divided by the calculated time of flight.

(ii) **2 points:** Award one point for a partially correct description, two points for a complete and correct description.

"Elastic" means that the total kinetic energy of the two carts was the same before and after collision. Before the collision, the only kinetic energy is that of Cart A: $\frac{1}{2}m_A v_A^2$. After the collision, the total kinetic energy is the sum of the kinetic energy of both carts, where each cart's kinetic energy is given by $\frac{1}{2}mv^2$. Compare the total kinetic energy after collision to Cart A's kinetic energy before collision. If these values are equal, the collision was elastic. If the kinetic energy after the collision is less than the kinetic energy before collision, the collision was *not* elastic.

Question 2

Part (a)

2 points: Award one point for a partially correct description, two points for a complete and correct description. Consider the center of the meterstick as the fulcrum; then the weight of the meterstick provides no torque. The oppositely directed torques applied by each scale must be equal, because the meterstick is in equilibrium. Torque is force times distance from the fulcrum; since the right-hand scale's torque calculation includes a smaller distance from the fulcrum, the right-hand scale must apply more force in order to multiply to the same torque.

Part (b)

4 points: Full credit for a complete and correct answer. Award three points partial credit for a correct approach with incorrect answers. Award two points for a correct approach and correct answer for one of the scales, but not the other. Award at least one point if the answer involved some use of torque equilibrium.

For this calculation, consider the left-hand scale as the fulcrum—that way, the left-hand scale provides no torque, and we only have to solve for one unknown variable. Set counter-clockwise torques equal to clockwise torques, with T_2 the reading in the right-hand scale.

The weight of the meterstick provides the clockwise torque; the right-hand scale provides the counterclockwise torque.

$$** \ (T_2)(50 \text{ cm}) = (1.5 \text{ N})(30 \text{ cm})$$

Solve for T_2 to get $T_2 = 0.9$ N
Next, the sum of the scale readings has to be the 1.5 N weight of the meterstick:

$$0.9 \text{ N} + T_1 = 1.5 \text{ N}$$

Giving $T_1 = 0.6$ N

Part (c)

(i) **2 points:** Award one point for a partially correct description, two points for a complete and correct description.

Look at the starred calculation in Part (b). By moving the right-hand scale closer to the center, the scale will be less than 50 cm from the left-hand scale; but the meterstick's center will still be 30 cm from the fulcrum. So when we solve for T_2, we're dividing (1.5 N) (30 cm) by a smaller value, giving a bigger T_2 reading.

But the question asks for the reading in the left-hand scale, which adds to T_2 to the same 1.5 N. A bigger T_2 adds to a smaller T_1 to get 1.5 N. Answer: decrease.

(ii) **2 points:** Award one point for a partially correct description, two points for a complete and correct description.

See Part (i): T_2, the reading in the right-hand scale, will increase.

Part (d)

2 points: Award one point for a partially correct description, two points for a complete and correct description.

Again, start from the equilibrium of torques using the left-hand scale as the fulcrum:

$$(T_2)(50 \text{ cm}) = (1.5 \text{ N})(30 \text{ cm})$$

Hanging a 0.2-N weight would provide a clockwise torque that would add to the torque applied by the meterstick's weight on the right of this equation. Algebraically, T_2 is increased by adding to the numerator of the right side of this equation. We want to add the biggest possible torque.

Torque is force times distance from the fulcrum. We want, then, the largest possible distance from the fulcrum, which would be the right-hand edge of the meterstick, 80 cm from the left-hand scale.

Question 3

Part (a)
1 point for *both* a correct answer *and* a correct justification.

Probe A. The probe's speed is the circumference of its circular motion divided by its period of revolution. We already established that the period is the same for each. Probe A has a bigger orbital radius, meaning a larger circumference of its circular motion, meaning a greater speed.

Part (b)
2 points: Award one point for the correct answer with a partially correct justification; award both points for the fully correct answer and justification.

Probe A. The centripetal acceleration is $\dfrac{v^2}{r}$. The problem is that Probe A has both a larger speed v and a larger orbital radius r. In order to answer the question, it's necessary to replace the speed v by circumference over period, $v = \dfrac{2\pi r}{T}$. Now the acceleration is $\dfrac{\left(\dfrac{2\pi r}{T}\right)^2}{r} = \dfrac{4\pi^2 r}{T^2}$. Okay, now we know: both probes have the same orbital period T, and r is in the numerator. The bigger-radius orbit—Probe A—has the greater acceleration.

Part (c)

(i) **1 point** for correct ranking

$$\textit{Greatest} \qquad 2 = 4 > 1 = 3 > 5 = 6 \qquad \textit{Least}$$

(ii) **3 points:** Award one point for justifying all three sets of force pairs set equal. Award one more point for justifying act least one correct portion of the ranking. Award the third point for justifying a second correct portion of the ranking.

By Newton's Third Law, the three force pairs can be immediately set equal: that's #1 with #3, #2 with #4, and #5 with #6. Next, we know that the force of Jupiter on either probe is given by $F = G\dfrac{Mm}{r^2}$, where M and m are the masses of Jupiter and the probe, respectively. Since the probes are identical, the numerator is the same for both #1 and #2, but the distance of the probe from Jupiter's center is smaller for Probe B. Therefore, Probe B experiences more force, and force #2 is greater than force #1. As for force #5, Jupiter is an enormous planet, many times more massive than Earth, even. There's no way that the product of the space probes' masses can ever approach Jupiter's mass, meaning that the numerator of the force equation must be way smaller for force #5.

Question 4

The paragraph response must discuss kinetic energy, total mechanical energy, and linear momentum for each of the two systems. For each of these three quantities in each system, award one point for correctly explaining whether it is conserved and correctly justifying why it is or isn't conserved. For example:

1 point: In system A, kinetic energy is *not* conserved. When the blocks are released, Block A speeds up away from Block B. Kinetic energy depends on mass and speed only. Since Block A's speed increases without changing its mass, kinetic energy cannot remain constant.

1 point: In system A, total mechanical energy is *not* conserved. Since the system consists only of Block A, there is no interaction with another object that would allow for the storage of potential energy. The force of the spring on Block A would be a force external to the system, and the spring does work on Block A because Block A moves parallel to the spring force; when a net force external to the system does work, mechanical energy is not conserved.

1 point: In system A, linear momentum is *not* conserved. Either the reasoning for system A's kinetic energy or total mechanical energy can be extended here. Linear momentum depends on mass and speed and Block A's speed changes without changing mass. Or, the spring force is external to the system, and momentum is only conserved in systems for which no net external force acts.

1 point: In system B, kinetic energy is *not* conserved. Kinetic energy is a scalar, so kinetic energy of a system of objects is just the addition of the kinetic energies of all the objects in the system. Both blocks speed up, so both blocks are increasing their kinetic energy, increasing the system's kinetic energy.

1 point: In system B, total mechanical energy *is* conserved. No force external to the spring-blocks system does work, so mechanical energy is conserved. The kinetic energy gained by the blocks was converted from potential energy stored in the spring.

1 point: In system B, linear momentum *is* conserved. No force external to the spring-blocks system acts, so linear momentum is conserved. Here even though Block B gains linear momentum, momentum is a vector—its gain of momentum is canceled by the momentum gained by Block A in the opposite direction.

Add 1 point if the paragraph correctly states whether each quantity is conserved in each system, regardless of whether the justifications are legitimate.

Question 5

Part (a)

1 point for correctly identifying and justifying Bulb 1's current increase and **1 point** for correctly identifying and justifying Bulb 3's decreased current.

Initially, the circuit is just Bulbs 1 and 3 in series. When Bulb 2 is added, the voltage from the battery is unchanged. Yet the total resistance of the circuit decreases, because an additional parallel path is added. Therefore, by $V = IR$ with constant V, the total current in the circuit increases.

Bulb 1 takes the total current, so Bulb 1's current increases. For Bulb 1 *only*, the resistance is a property of the bulb and thus doesn't change. So by $V = IR$ with constant R, Bulb 1 takes an increased voltage, too.

Then by Kirchoff's loop rule, an increase voltage across Bulb 1 means a decreased voltage across Bulb 3. And for Bulb 3 only, by $V = IR$ with constant R, Bulb 3's current also decreases. (Obviously Bulb 2's current increases from nothing to something.)

Part (b)

1 point for either a correct answer with correct justification; or, for an answer consistent with the answers to Part (a) with reference to the power dissipated by the bulbs.

All bulbs have an unchanging resistance. Brightness depends on power, which is I^2R. With constant R, a bigger current means more brightness; a smaller current means less brightness. So Bulb 1 gets brighter and Bulb 3 gets dimmer.

Part (c)

2 points for a fully correct answer with justification. One of these two points can be earned for a partially correct justification, or for an incorrect answer that is justified consistently with the answers to (A) or (B).

For the whole circuit, use power = V^2/R. The voltage of the battery is unchanged because it's still the same battery. The resistance of the circuit decreases because of the extra parallel path. So, decreasing the denominator increases the entire value of the equation, so power increases.

Part (d)

2 points for a complete and correct explanation. One of these two points can be earned by a partially correct, or an incomplete, justification.

Conservation of charge in circuits is expressed in Kirchoff's junction rule—the current entering a junction equals the current leaving the junction. At any given moment of time, the junction rule holds. Now, when the switch is closed, more current flows from the battery than before. That's not a violation of charge conservation, because the materials in the battery contain way more charged particles than are ever flowing through the wires. After the switch is closed, more current flows into the junction right before the switch than before, but more current *also* flows *out* of that junction than before. Charge conservation doesn't mean that the same current must always flow in a circuit, it just says that whatever charge does flow in a circuit must flow along the wires.

Practice Exam 2

ANSWER SHEET FOR SECTION I

1 (A) (B) (C) (D)
2 (A) (B) (C) (D)
3 (A) (B) (C) (D)
4 (A) (B) (C) (D)
5 (A) (B) (C) (D)
6 (A) (B) (C) (D)
7 (A) (B) (C) (D)
8 (A) (B) (C) (D)
9 (A) (B) (C) (D)
10 (A) (B) (C) (D)
11 (A) (B) (C) (D)
12 (A) (B) (C) (D)
13 (A) (B) (C) (D)
14 (A) (B) (C) (D)
15 (A) (B) (C) (D)
16 (A) (B) (C) (D)
17 (A) (B) (C) (D)

18 (A) (B) (C) (D)
19 (A) (B) (C) (D)
20 (A) (B) (C) (D)
21 (A) (B) (C) (D)
22 (A) (B) (C) (D)
23 (A) (B) (C) (D)
24 (A) (B) (C) (D)
25 (A) (B) (C) (D)
26 (A) (B) (C) (D)
27 (A) (B) (C) (D)
28 (A) (B) (C) (D)
29 (A) (B) (C) (D)
30 (A) (B) (C) (D)
31 (A) (B) (C) (D)
32 (A) (B) (C) (D)
33 (A) (B) (C) (D)
34 (A) (B) (C) (D)

35 (A) (B) (C) (D)
36 (A) (B) (C) (D)
37 (A) (B) (C) (D)
38 (A) (B) (C) (D)
39 (A) (B) (C) (D)
40 (A) (B) (C) (D)
41 (A) (B) (C) (D)
42 (A) (B) (C) (D)
43 (A) (B) (C) (D)
44 (A) (B) (C) (D)
45 (A) (B) (C) (D)
46 (A) (B) (C) (D)
47 (A) (B) (C) (D)
48 (A) (B) (C) (D)
49 (A) (B) (C) (D)
50 (A) (B) (C) (D)

$$V = \sqrt{\dfrac{T}{\frac{m}{L}}} \qquad V = f\lambda$$

AP Physics 1 Practice Exam 2: Section I (Multiple-Choice)

Directions: The multiple-choice section consists of 50 questions to be answered in **90 minutes**. You may write scratch work in the test booklet itself, but only the answers on the answer sheet will be scored. You may use a calculator, the equation sheet, and the table of information.

Questions 1–45: Single-Choice Items

Directions: Choose the single best answer from the four choices provided and grid the answer with a pencil on the answer sheet.

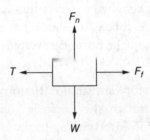

Questions 1 and 2 refer to the following information: A person pulls on a string, causing a block to move to the left at a constant speed. The free body diagram shows the four forces acting on the block: the tension (T) in the string, the normal force (F_n), the weight (W), and the friction force (F_f). The coefficient of friction between the block and the table is 0.30.

1. Which is the Newton's third law force pair to T?

 (A) The force of the block on the string
 (B) The force of the block on the table
 (C) The force of the table on the block
 (D) The force of friction on the block

2. Which of the following correctly ranks the four forces shown?

 (A) $T > F_f > W = F_n$
 (B) $W = F_n > T = F_f$
 (C) $W = F_n > T > F_f$
 (D) $W = F_n = T = F_f$

3. Which of the following circuits shows a placement of meters and an observation of their readings that would allow researchers to experimentally

demonstrate whether energy is conserved in the circuit?

(A) Researchers look to see whether the readings on voltmeters 1 and 2 add to the reading in voltmeter 3.

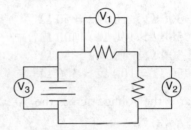

(B) Researchers look to see whether the voltmeter readings are equal.

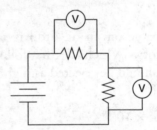

(C) Researchers look to see whether the readings on ammeters 1 and 2 add to the reading in ammeter 3.

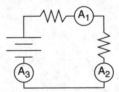

(D) Researchers look to see whether the ammeter readings are equal.

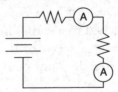

GO ON TO THE NEXT PAGE

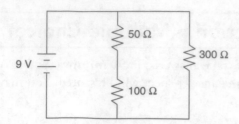

Questions 4 and 5 refer to the circuit shown in the figure, which includes a 9 V battery and three resistors.

4. Which of the following ranks the resistors by the charge that flows through each in a given time interval?

   (A) $300\ \Omega > 100\ \Omega = 50\ \Omega$
   (B) $50\ \Omega > 100\ \Omega > 300\ \Omega$
   (C) $300\ \Omega > 100\ \Omega > 50\ \Omega$
   (D) $50\ \Omega = 100\ \Omega > 300\ \Omega$

5. What is the voltage across the 50 Ω resistor?

   (A) 9.0 V
   (B) 6.0 V
   (C) 3.0 V
   (D) 1.0 V

6. The charge on an oil drop is measured in the laboratory. Which of the following measurements should be rejected as highly unlikely to be correct?

   (A) 6.4×10^{-19} C
   (B) 8.0×10^{-19} C
   (C) 4.8×10^{-19} C
   (D) 2.4×10^{-19} C

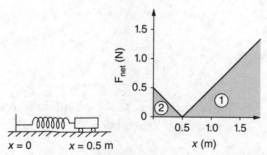

7. A cart attached to a spring initially moves in the x direction at a speed of 0.40 m/s. The spring is neither stretched nor compressed at the cart's initial position ($x = 0.5$ m). The figure shows a graph of the magnitude of the net force experienced by the cart as a function of x, with two areas under the graph labeled. Is it possible to analyze the graph to determine the change in the cart's kinetic energy as it moves from its initial position to $x = 1.5$ m?

   (A) No, the cart's mass must be known.
   (B) Yes, subtract area 2 from area 1.
   (C) Yes, add area 2 to area 1.
   (D) Yes, determine area 1.

8. A horse is attached to a cart that is at rest behind it. Which force, or combination of forces, explains how the horse-cart system can accelerate from rest?

   (A) The forward static friction force of the ground on the horse is greater than any friction forces acting backward on the cart, providing a forward acceleration.
   (B) The forward force of the horse on the cart is greater than the backward force of the cart on the horse, providing a forward acceleration.
   (C) The force of the horse's muscles on the rest of the horse-cart system provides the necessary acceleration.
   (D) The upward normal force of the ground on the horse is greater than the horse's weight, providing an upward acceleration.

9. A pipe full of air is closed at one end. A standing wave is produced in the pipe, causing the pipe to sound a note. Which of the following is a correct statement about the wave's properties at the closed end of the pipe?

   (A) The pressure is at a node, but the particle displacement is at an antinode.
   (B) The pressure is at an antinode, but the particle displacement is at a node.
   (C) The pressure and the particle displacement are both at nodes.
   (D) The pressure and the particle displacement are both at antinodes.

GO ON TO THE NEXT PAGE

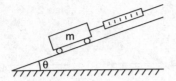

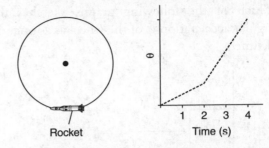

Rocket   Time (s)

10. In the laboratory, a cart of mass *m* is held in place on a smooth incline by a rope attached to a spring scale, as shown in the figure. The angle of the incline from the horizontal θ varies between 0° and 90°. A graph of the reading in the spring scale as a function of the angle θ is produced. Which of the following will this graph look like?

(A)
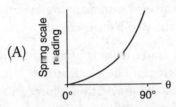

(B)
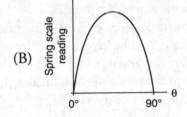

(C)
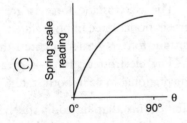

(D)
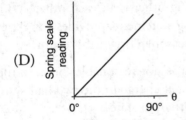

Questions 11 and 12 refer to the following information:
A bicycle wheel of known rotational inertia is mounted so that it rotates clockwise around a vertical axis, as shown in the first figure. Attached to the wheel's edge is a rocket engine, which applies a clockwise torque τ on the wheel for a duration of 0.10 s as it burns. A plot of the angular position θ of the wheel as a function of time *t* is shown in the graph.

11. In addition to the wheel's rotational inertia and the duration of time the engine burns, which of the following information from the graph would allow determination of the net torque the rocket exerts on the wheel?

(A) The area under the graph between $t = 0$ s and $t = 3$ s

(B) The change in the graph's slope before and after $t = 2$ s

(C) The vertical axis reading of the graph at $t = 3$ s

(D) The vertical axis reading of the graph at $t = 2$ s

12. Which of the following graphs sketches the angular acceleration α of the wheel as a function of time?

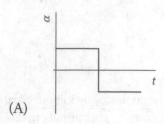

(A)

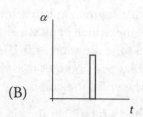

(B)

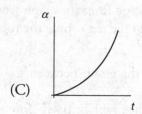

(C)

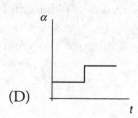

(D)

13. A rock drops onto a pond with a smooth surface. A few moments later, the wave produced by the rock's impact arrives at the shore, touching the ankles of a wading child. Which of the following observations provides evidence that the portion of the wave hitting the child's ankles carries less energy than the wave did when it was first created by the rock's impact?

(A) The wave is moving more slowly.
(B) The wave pulse's width has become greater.
(C) The wave pulse's width has become smaller.
(D) The wave's height has become smaller.

State 1

State 2

Questions 14 and 15 refer to the following information:
Two metal balls of equal mass (100 g) are separated by a distance (d). In state 1, shown in the figure, the left ball has a charge of +20 µC, while the right ball has a charge of –20 µC. In state 2, both balls have an identical charge of –40 µC.

14. Which of the following statements about the magnitudes of the electrostatic and gravitational forces between the two balls is correct?

(A) The electrostatic force is greater in state 2 than in state 1. The electrostatic force is greater than the gravitational force in state 1, but the gravitational force is greater than the electrostatic force in state 2.
(B) The electrostatic force is greater in state 1 than in state 2. The electrostatic force is greater than the gravitational force in state 1, but the gravitational force is greater than the electrostatic force in state 2.
(C) The electrostatic force is greater in state 2 than in state 1. The electrostatic force is greater than the gravitational force in both states.
(D) The electrostatic force is greater in state 1 than in state 2. The electrostatic force is greater than the gravitational force in both states.

15. An experimenter claims that he took the balls from state 1 to state 2 without causing them to contact anything other than each other. Which of the following statements provides correct evidence for the reasonability of this claim?

(A) The claim is not reasonable, because there was more net charge in the two-ball system in state 2 than in state 1.
(B) The claim is not reasonable, because there was identical charge on each ball in state 2.
(C) The claim is reasonable, because in both states, each ball carried the same magnitude of charge as the other.
(D) The claim is reasonable, because the positive charge in state 1 could have been canceled out by the available negative charge.

GO ON TO THE NEXT PAGE

16. A guitar string creates a sound wave of known frequency. Which of the following describes a correct and practical method of measuring the wavelength of the sound wave with a meterstick?

    (A) Lightly touch the guitar string in the middle such that a single node is created. Measure the length of the string; this is the wavelength.

    (B) Measure the length of the guitar string; this is half the wavelength.

    (C) Adjust the length of a pipe placed near the string so that resonances are heard. Measure the difference between the pipe lengths for consecutive resonances; this is half the wavelength.

    (D) Measure the peak-to-peak distance of the wave as it passes; this is the wavelength.

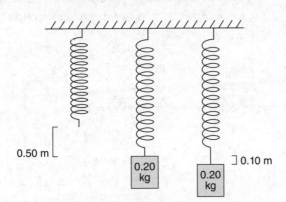

17. When a 0.20 kg block hangs at rest vertically from a spring of force constant 4 N/m, the spring stretches 0.50 m from its unstretched position, as shown in the figure. Subsequently, the block is stretched an additional 0.10 m and released such that it undergoes simple harmonic motion. What is the maximum kinetic energy of the block in its harmonic motion?

    (A) 0.50 J
    (B) 0.02 J
    (C) 0.72 J
    (D) 0.20 J

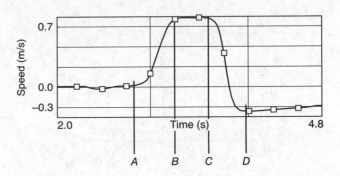

Questions 18 and 19 refer to the following information:

A student pushes cart A toward a stationary cart B, causing a collision. The velocity of cart A as a function of time is measured by a sonic motion detector, with the resulting graph shown in the figure.

18. At which labeled time did the collision begin to occur?

    (A) A
    (B) B
    (C) C
    (D) D

19. What additional measurements, in combination with the information provided in the graph, could be used to verify that momentum was conserved in this collision?

    (A) The mass of each cart and cart B's speed after the collision

    (B) The force of cart A on cart B, and cart B's speed after the collision

    (C) The mass of each cart only

    (D) The force of cart A on cart B only

20. A car of mass m initially travels at speed v. The car brakes to a stop on a road that slants downhill, such that the car's center of mass ends up a vertical height h below its position at the start of braking. Which of the following is a correct expression for the increase in the internal energy of the road-car system during the braking process?

    (A) $\frac{1}{2}mv^2 - mgh$
    (B) $\frac{1}{2}mv^2$
    (C) 0
    (D) $\frac{1}{2}mv^2 + mgh$

GO ON TO THE NEXT PAGE

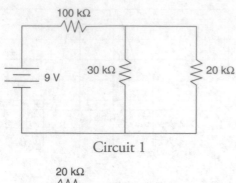

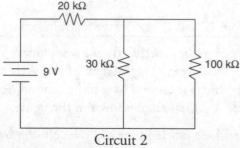

21. The circuits shown in the figure contain the same three resistors connected in different configurations. Which of the following correctly explains which configuration (if either) produces the larger current coming from the battery?

(A) Circuit 1; the larger resistor is closer to the battery.
(B) Circuit 2; the equivalent resistance of the three resistors is smaller than in circuit 1.
(C) Neither; the batteries are both the same.
(D) Neither; the individual resistors are the same in each circuit, even if they are in a different order.

22. A student pushes a puck across a table, moving it from position $x = 0$ to position $x = 0.2$ m. After he lets go, the puck continues to travel across the table, coming to rest at position $x = 1.2$ m. When the puck is at position $x = 1.0$ m, which of the following is a correct assertion about the net force on the puck?

(A) The net force is in the negative direction, because the puck is moving in the positive direction but slowing down.
(B) The net force is down, because the puck is near the Earth, where gravitational acceleration is 10 m/s² downward.
(C) The net force is in the positive direction, because the student's push caused the puck to speed up in the positive direction.
(D) The net force is zero, because the student's push in the positive direction must equal the force of friction in the negative direction.

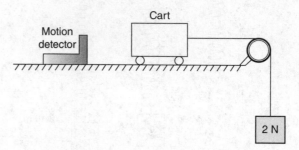

23. In an experiment, a cart is placed on a flat, negligible-friction track. A light string passes over a nearly ideal pulley. An object with a weight of 2.0 N hangs from the string. The system is released, and the sonic motion detector reads the cart's acceleration. Can this setup be used to determine the cart's inertial mass?

(A) Yes, by dividing 2.0 N by the acceleration, and then subtracting 0.2 kg.
(B) No, because only the cart's gravitational mass could be determined.
(C) Yes, by dividing 2.0 N by the acceleration.
(D) No, because the string will have different tensions on either side of the pulley.

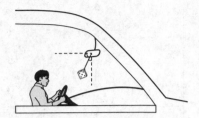

24. An object hangs by a string from a car's rear-view mirror, as shown in the figure. The car is speeding up and moving to the right. Which of the following diagrams correctly represents the forces acting on the object?

(A) T: Force of string on object
W: Force of Earth on object
F_o: Force of object on string

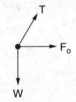

(B) T: Force of string on object
W: Force of Earth on object
F_o: Force of object on string

(C) T: Force of string on object
W: Force of Earth on object
F_o: Force of object on string

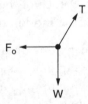

(D) T: Force of string on object
W: Force of Earth on object

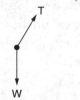

25. The diagram shown here represents the particles in a longitudinal standing wave. Which of the following is an approximate measure of the standing wave's maximum amplitude?

(A) Half the distance from 1 to 4
(B) The distance from 2 to 3
(C) The distance from 1 to 4
(D) Half the distance from 2 to 3

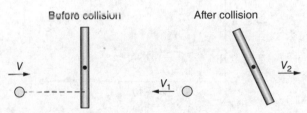

Questions 26 and 27 refer to the following information:
A small ball of mass m moving to the right at speed v collides with a stationary rod, as shown in the figure. After the collision, the ball rebounds to the left with speed v_1, while the rod's center of mass moves to the left at speed v_2. The rod also rotates counterclockwise.

26. Which of the following equations determines the rod's change in angular momentum about its center of mass during the collision?

(A) $I\omega$ where I is the rod's rotational inertia about its center of mass, and ω is its angular speed after collision.
(B) Iv_2/r, where I is the rod's rotational inertia about its center of mass, and r is half the length of the rod.
(C) mv_1d, where d is the distance between the line of the ball's motion and the rod's center of mass.
(D) mv_1r, where r is half the length of the rod.

GO ON TO THE NEXT PAGE

27. Is angular momentum about the rod's center of mass conserved in this collision?

    (A) No, the ball always moves in a straight line and thus does not have angular momentum.
    (B) No, nothing is spinning clockwise after the collision to cancel the rod's spin.
    (C) Yes, the only torques acting are the ball on the rod and the rod on the ball.
    (D) Yes, the rebounding ball means the collision was elastic.

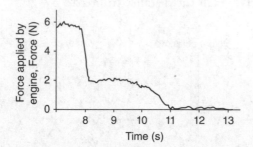

Questions 28 and 29 refer to the following information: A model rocket with a mass of 100 g is launched straight up. Eight seconds after launch, when it is moving upward at 110 m/s, the force of the engine drops as shown in the force-time graph.

28. Which of the following is the best estimate of the impulse applied by the engine to the rocket after the $t = 8$ s mark?

    (A) 100 N·s
    (B) 20 N·s
    (C) 5 N·s
    (D) 50 N·s

29. Which of the following describes the motion of the rocket between $t = 8$ s and $t = 10$ s?

    (A) The rocket moves upward and slows down.
    (B) The rocket moves downward and speeds up.
    (C) The rocket moves upward at a constant speed.
    (D) The rocket moves upward and speeds up.

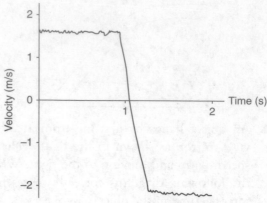

30. The velocity-time graph shown here represents the motion of a 500 g cart that initially moved to the right along a track. It collided with a wall at approximately time (t) = 1.0 s. Which of the following is the best estimate of the impulse experienced by the cart in this collision?

    (A) 3.6 N·s
    (B) 0.5 N·s
    (C) 0.2 N·s
    (D) 1.8 N·s

31. The Space Shuttle orbits 300 km above Earth's surface; Earth's radius is 6,400 km. What is the gravitational acceleration experienced by the Space Shuttle?

    (A) Zero
    (B) 4.9 m/s^2
    (C) 9.8 m/s^2
    (D) 8.9 m/s^2

32. A person stands on a scale in an elevator. He notices that the scale reading is less than his usual weight. Which of the following could possibly describe the motion of the elevator?

    (A) It is moving downward and slowing down.
    (B) It is moving upward and slowing down.
    (C) It is moving upward at a constant speed.
    (D) It is moving downward at a constant speed.

33. A textbook weighs 30 N at sea level. Earth's radius is 6,400 km. Which of the following is the best estimate of the textbook's weight on a mountain peak located 6,000 m above sea level?

    (A) 60 N
    (B) 15 N
    (C) 30 N
    (D) 7.5 N

GO ON TO THE NEXT PAGE

34. A satellite orbits the moon in a circle of radius R. For the satellite to double its speed but maintain a circular orbit, what must the new radius of its orbit be?

(A) ½R
(B) $4R$
(C) ¼R
(D) $2R$

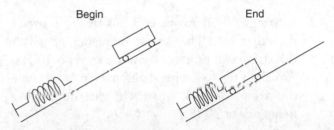

Begin   End

35. A 0.5 kg cart begins at rest at the top of an incline, 0.06 m vertically above its end position. It is released and allowed to travel down the smooth incline, where it compresses a spring. Between the positions labeled "Begin" and "End" in the figure, the work done on the cart by the earth is 0.30 J; the work done on the cart by the spring is –0.20 J. What is the cart's kinetic energy at the position labeled "End"?

(A) 0.80 J
(B) 0.10 J
(C) 0.50 J
(D) 0.40 J

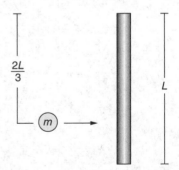

Questions 36 and 37 refer to the following information:
A rigid rod of length L and mass M sits at rest on an air table with negligible friction. A small blob of putty with a mass of m moves to the right on the same table, as shown in overhead view in the figure. The putty hits and sticks to the rod, a distance of $2L/3$ from the top end.

36. How will the rod-putty system move after the collision?

(A) The system will have no translational motion, but it will rotate about the rod's center of mass.
(B) The system will move to the right and rotate about the rod-putty system's center of mass.
(C) The system will move to the right and rotate about the rod's center of mass.
(D) The system will have no translational motion, but it will rotate about the rod-putty system's center of mass.

37. What quantities, if any, must be conserved in this collision?

(A) Linear momentum only
(B) Neither linear nor angular momentum
(C) Angular momentum only
(D) Linear and angular momentum

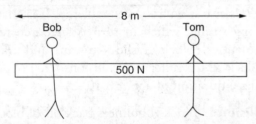

38. Bob and Tom hold a rod with a length of 8 m and weight of 500 N. Initially, Bob and Tom each hold the rod 2 m from the its ends, as shown in the figure. Next, Tom moves slowly toward the right edge of the rod, maintaining his hold. As Tom moves to the right, what happens to the torque about the rod's midpoint exerted by each person?

(A) Bob's torque decreases, and Tom's torque increases.
(B) Bob's torque increases, and Tom's torque decreases.
(C) Both Bob's and Tom's torque increases.
(D) Both Bob's and Tom's torque decreases.

GO ON TO THE NEXT PAGE

39. An object of mass m hangs from two ropes at unequal angles, as shown in the figure. Which of the following makes correct comparisons between the horizontal and vertical components of the tension in each rope?

(A) Horizontal tension is equal in both ropes, but vertical tension is greater in rope A.

(B) Both horizontal and vertical tension are equal in both ropes

(C) Horizontal tension is greater in rope B, but vertical tension is equal in both ropes.

(D) Both horizontal and vertical tension are greater in rope B

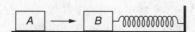

Questions 40 and 41 refer to the following information: Block B is at rest on a smooth tabletop. It is attached to a long spring, which in turn is anchored to the wall. Identical block A slides toward and collides with block B. Consider two collisions, each of which occupies a duration of about 0.10 s:

Collision I: Block A bounces back off of block B.
Collision II: Block A sticks to block B.

40. In which collision, if either, does block B move faster immediately after the collision?

(A) In collision I, because block A experiences a larger change in momentum, and conservation of momentum requires that block B does as well.

(B) In collision I, because block A experiences a larger change in kinetic energy, and conservation of energy requires that block B does as well.

(C) In neither collision, because conservation of momentum requires that both blocks must have the same momentum as each other in each collision.

(D) In neither collision, because conservation of momentum requires that both blocks must change their momentum by the same amount in each collision.

41. In which collision, if either, is the period and frequency of the ensuing oscillations after the collision larger?

(A) Period and frequency are the same in both.

(B) Period is greater in collision II, and frequency is greater in collision I.

(C) Period and frequency are both greater in collision I.

(D) Period and frequency are both greater in collision II.

42. A string is fixed at one end but free to move at the other end. The lowest frequency at which this string will produce standing waves is 10 Hz. Which of the following diagrams represents how the string will look when it is vibrating with a frequency of 30 Hz?

(A)

(B)

(C)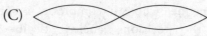

(D)

GO ON TO THE NEXT PAGE

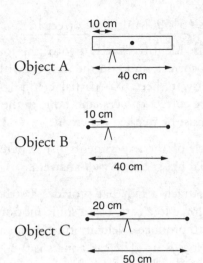

Object A
10 cm
40 cm

Object B
10 cm
40 cm

Object C
20 cm
50 cm

Stationary source

43. Three equal-mass objects (A, B, and C) are each initially at rest horizontally on a pivot, as shown in the figure. Object A is a 40 cm long, uniform rod, pivoted 10 cm from its left edge. Object B consists of two heavy blocks connected by a very light rod. It is also 40 cm long and pivoted 10 cm from its left edge. Object C consists of two heavy blocks connected by a very light rod that is 50 cm long and pivoted 20 cm from its left edge. Which of the following correctly ranks the objects' angular acceleration about the pivot point when they are released?

(A) A = B > C
(B) A > B = C
(C) A < B < C
(D) A > B > C

44. A stationary wave source emits waves at a constant frequency. As these waves move to the right, they are represented by the wave front diagram shown here. Sometime later, the wave source is moving at a constant speed to the right. Which of the following wave front diagrams could represent the propagation of the waves produced by the source?

Motion of source

(A)

(B)

(C)

(D)

45. An object of mass m is attached to an object of mass $3m$ by a rigid bar of negligible mass and length L. Initially, the smaller object is at rest directly above the larger object, as shown in the figure. How much work is necessary to flip the object 180°, such that the larger mass is at rest directly above the smaller mass?

(A) $2\pi mgL$

(B) $4mgL$

(C) $4\pi mgL$

(D) $2mgL$

Questions 46–50: Multiple-Correct Items

Directions: Identify **exactly two** of the four answer choices as correct, and grid the answers with a pencil on the answer sheet. No partial credit is awarded; both of the correct choices, and none of the incorrect choices, must be marked for credit.

46. Which of the experiments listed here measure inertial mass? **Select two answers.**

(A) Attach a fan that provides a steady 0.2 N force to a cart. Use a sonic motion detector to produce a velocity-time graph as the cart speeds up. The cart's mass is 0.2 N divided by the slope of the velocity-time graph.

(B) Hang an object from a spring and cause the object to oscillate in simple harmonic motion. Use a stopwatch to determine the period of the motion. The object's mass is this period squared, divided by $4\pi^2$, divided by the spring's force constant.

(C) Place an object on one side of a two-pan balance scale. On the other pan, place objects with previously calibrated masses. The object's mass is equal to the amount of calibrated mass on the other plate when the plates balance.

(D) Hang a spring from a clamp, then hook an object on the spring. Use a ruler to measure how far the spring stretched once the object was attached. This distance multiplied by the force constant of the spring determines the object's weight. Use 10 N/kg to convert this weight to mass.

47. An object on a spring vibrates in simple harmonic motion. A sonic motion detector is placed under the object. Which of the following determines the period of the object's oscillation? **Select two answers.**

(A) The maximum slope on a position-time graph divided by the maximum slope on a velocity-time graph

(B) The maximum vertical axis value on a position-time graph divided by the maximum vertical axis value on a velocity-time graph

(C) The time between positive maxima on a velocity-time graph

(D) The time between positive maxima on a position-time graph

48. Two carts on a negligible-friction surface collide with each other. Which of the following is a correct statement about an elastic collision between the carts? **Select two answers.**

(A) Some of the kinetic energy of the two-cart system is converted to thermal and sound energy.
(B) The carts bounce off of each other.
(C) Linear momentum is conserved.
(D) The carts stick together.

49. The resistance of a sample of circular cross-section wire is known. A measurement of which two of the following would allow for a calculation of the resistivity of the wire? **Select two answers.**

(A) The wire's diameter
(B) The wire's density
(C) The wire's length
(D) The wire's temperature

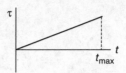

50. The net torque τ on an object of rotational inertia I is shown as a function of time t. At time t_{max}, the object has speed ω and angular acceleration α. Which of the following methods correctly determines the change in the object's angular momentum? **Select two answers.**

(A) Multiply I by α/t_{max}.
(B) Multiply the average torque by t_{max}.
(C) Calculate the area under the line on the graph.
(D) Multiply I by ω.

STOP. End of Physics 1 Practice Exam 2—Multiple-Choice Questions

AP Physics 1 Practice Exam 2: Section II (Free-Response)

Directions: The free-response section consists of five questions to be answered in **90 minutes**. Budget approximately 20 to 25 minutes each for the first two longer questions; the next three shorter questions should take about 12 to 17 minutes each. Explain all solutions thoroughly, as partial credit is available. On the actual test, you will write the answers in the test booklet; for this practice exam, you will need to write your answers on a separate sheet of paper.

1. (7 points)

In the laboratory, you are given three resistors: $R_1 = 30$ kΩ, $R_2 = 60$ kΩ, and $R_3 = 120$ kΩ. You are to use these three resistors in a circuit with a 12 V battery such that one resistor is in series with the battery, and the other two are parallel with each other.

(a) Diagram a circuit with these three resistors such that the resistor in series with the battery takes the largest possible voltage across it. Justify your answer.

(b) For these three given resistors, is it possible for the resistor in series with the battery to take a smaller voltage than either of the two parallel resistors? Justify your answer.

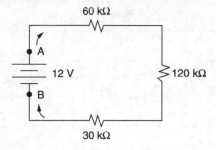

(c) Now the resistors are connected in series with the battery, as shown in the figure. On the axes provided, sketch a graph of the electric potential (V) measured along the circuit, starting at position A, going in the direction shown, and ending at position B. Consider the electric potential at position B to be zero.

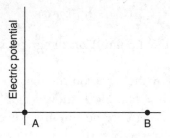

GO ON TO THE NEXT PAGE

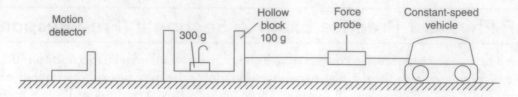

2. (12 points)

In the laboratory, a student connects a toy vehicle to a hollow block with a string. The hollow block has a mass of 100 g and contains an additional 300 g object. The vehicle is turned on, causing it to move forward along a table at a constant speed. A force probe records the tension in the string as a function of time, and a sonic motion detector reads the position of the cart as a function of time. The positioning of the probes is shown in the diagram. The data collected are shown below.

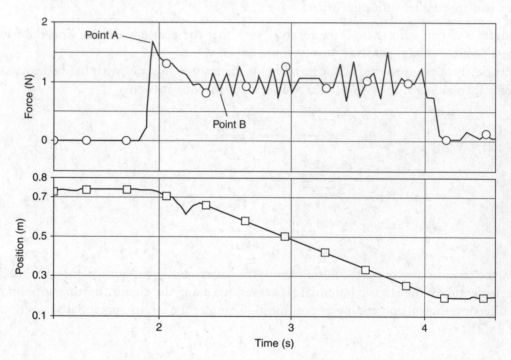

(a) Explain why the force reading marked point A on the graph is significantly different from the reading marked point B on the graph

(b) i. Calculate the impulse applied by the string on the block.

 ii. A student in the lab contends, "The block moved at a constant speed, so it has no change in momentum and should thus experience no impulse." Evaluate the validity of this student's statement with reference to the answer to part (i).

Now the student is asked to determine whether the coefficient of kinetic friction between the block and the table depends on the block's speed.

(c) Describe an experimental procedure that the student could use to collect the necessary data, including all the equipment he or she would need.

(d) How should the student analyze the data to determine whether the coefficient of friction depends on the block's speed? What evidence from the analysis would be used to make the determination?

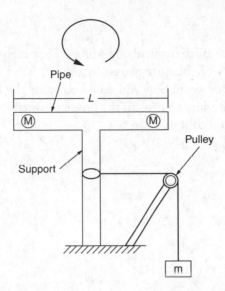

3. (12 points)

An object of mass *m* is attached via a rope to the stem of the device, as shown in the figure. The top portion of the device includes two rocks, each with a mass of *M*, located near the end of a hollow horizontal pipe of length *L*. The rotational inertia of the pipe itself and the cylindrical support is assumed to be negligible compared to that of the rocks inside the pipe. As the object falls from rest, the device begins to rotate.

(a) Explain why the tension in the hanging rope is not equal to *mg*.

(b) Can the angular acceleration of the device be calculated given the tension in the rope *T* and the other information provided in the description? If so, explain in several sentences how the calculation could be performed. If not, explain in several sentences why not, including what additional measurements would be necessary and how those measurements could be performed. In either case, you should not actually do the calculations, but provide complete instructions so that another student could use them.

(c) Derive an expression for the rotational inertia of the device. If you need to define new variables to represent easily measurable quantities, do so clearly.

(d) How would replacing the hanging object with a new object with a mass of *2m* affect the angular acceleration of the device? Answer with specific reference to the equation you derived in (c).

(e) Explain how the reasonability of the assumption that the rotational inertia of the pipe and the support is negligible in the calculation of the device's rotational inertia could be justified. You may use either a theoretical or an experimental approach to your justification.

4. (7 points)

A spring is attached vertically to a table. Undisturbed, the spring has a length of 20 cm. In procedure A, a block with a mass of 400 g is placed gently on top of the spring, compressing it so its length is 15 cm. In procedure B, a student pushes the block farther down, such that the spring has a length of 10 cm, and then releases the block from rest. The block is projected to height h.

   Consider a system consisting of the block, the spring, and Earth. Define the potential energy of this system as zero when the spring has a length of 15 cm.

(a) i. Calculate the work done by the student on the block-spring-Earth system in procedure B.
    ii. Justify your choice of equations in your calculation, as well as any distance values you used.

(b) Now the same spring is used with a block whose mass is 800 g. In procedure C, the 800 g block is placed gently on top of the spring, compressing it to a length of 10 cm. In procedure D, a student pushes the block 5 cm farther down and releases it from rest. In terms of *h*, how high will this 800 g block be projected? Justify your answer.

5. [paragraph response] (7 points)

Two speakers are set up outside, far away from any buildings. They each play steady-pitched notes of slightly different frequency. The frequency of the note played by the left-hand speaker is 350 Hz.

In a clear, coherent paragraph, describe how you could estimate the frequency in the right-hand speaker using no equipment whatsoever. Be sure to explain the physics principles that underlie your method, as well as the specific procedure you would follow.

Solutions: AP Physics 1 Practice Exam 2, Section I (Multiple-Choice)

Questions 1–45: Single-Choice Items

1. **A**—The tension is the force of the string acting on the block. Newton's third law says that since the string pulls on the block, the block pulls equally on the string. (The answer is NOT for force of friction: Newton's third law force pairs can never act on the same object.)

2. **B**—Since the block has no acceleration, left forces equal right forces, and up forces equal down forces. The equation for a force of friction is $F_f = \mu F_m$. Since the coefficient of friction μ is less than 1, the friction force must be less than the normal force.

3. **A**—The energy gained or lost by a charge as it goes through a circuit element is voltage, and a voltmeter measures voltage. In this circuit loop, any energy gained by charges across the battery must equal the energy lost by the charges through the two resistors; this is known as Kirchoff's Loop Rule.

4. **D**—"Charge that flows through each in a given time interval" is a complicated way of saying "current." Current through series resistors must always be the same through each, so the 50 Ω resistor and the 100 Ω resistor should rank equally. Then simplify the circuit to two parallel branches. The left branch has an equivalent resistance of 150 Ω and the right branch of 300 Ω. With the same voltage across each branch, the larger current goes through the path with smaller resistance by Ohm's law.

5. **C**—Consider just the two series resistors, which have an equivalent resistance of 150 Ω and are connected to the 9 V battery. The current through that branch of the circuit is $\frac{9\,\text{V}}{150\,\Omega} =$ 0.06 A. Now consider just the 50 Ω resistor. The voltage across it is (0.06 A)(50 Ω) = 3 V. As a sanity check, we know that for two resistors in series (which take the same current), $V = IR$ says that the smaller resistor takes less voltage; 3 V across the 50 Ω resistor leaves 6 V across the 100 Ω resistor.

6. **D**—All experimental evidence in existence shows that the smallest unit of isolated charge is the electron charge $e = 1.6 \times 10^{-19}$ C. The charge measured in A is equivalent to 4 fundamental charges; in B, 5 fundamental charges; and in C, 3 fundamental charges. But choice D would require an isolated charge of 1.5 e, which is not possible.

7. **D**—The work done by the net force is the area under the graph. Since the cart only moved from position $x = 0.5$ m to $x = 1.5$ m, area 1 is the work done by the net force. By the work-energy theorem, work done by the net force is the change in an object's kinetic energy. (Yes, the mass must be known to determine the values of the initial and final kinetic energy; however, the question asks only for the *change* in kinetic energy.)

8. **A**—To evaluate the acceleration of the horse-cart system, you can only consider the forces applied by objects external to the system. This eliminates choices B and C, which discuss forces of the system on itself. Choice D is ridiculous because the horse remains on the ground. Choice A is correct: the ground pushes forward on the horse's hooves because the horse's hooves push backward on the ground.

9. **B**—The closed end of the pipe makes it impossible for the air particles right next to the end to vibrate at all, which is why a particle displacement node must exist there. It's easiest, I think, just to remember that in a sound wave, pressure is at an antinode when particle displacement is at a node and vice versa. Or remember that the pressure must be equal to atmospheric pressure at the open end, which is exposed to the atmosphere, and greater inside because the air is compressed, not sucked out of the pipe.

10. **C**—The cart is in equilibrium, so the spring scale's force is equal to the component of the cart's weight that acts parallel to the plane. That component is $mg \sin\theta$. Thus, the graph of spring scale reading vs. θ should be a sine function, as in choice C. Choice B is wrong because that's what

a sine function looks like all the way to 180°; here, the angle of the incline only goes to 90°.

11. **B**—Newton's second law for rotation says $\tau_{net} = I\alpha$, where α is the angular acceleration, or change in the wheel's angular velocity per time. To find the wheel's angular velocity, look at the slope of the angular position versus time graph. The slope changes after the torque is applied; so the change in the slope is the change in the angular velocity, which (when divided by the 0.10 s duration of the rocket firing) gives the angular acceleration.

12. **B**—Angular acceleration is the change in angular speed in each second. To find angular speed, take the slope of the angular position versus time graph shown. This slope is constant for two seconds, then it changes to another constant slope. Therefore, there is no angular acceleration during the time when the slope doesn't change—with no change in angular speed, there's no acceleration. The only change in angular speed comes at the moment the slope changes, so that's the only time when there's any angular acceleration.

13. **D**—The energy carried by a wave depends on the wave's amplitude, meaning its height. The speed and pulse width or wavelength have no impact on the energy of a wave.

14. **C**—The magnitude of the electrostatic force depends on the product of the charges, regardless of sign. In state 2, the charges are both bigger than in state 1, giving a bigger electrostatic force in state 2. Without even doing the calculation, you can recognize that since Newton's gravitational constant G is orders of magnitude less than the coulomb's law constant k, the electrostatic force will be much greater than the gravitational force between two objects in virtually any laboratory situation. The only time these forces become comparable is when objects the size of planets exert gravitational forces.

15. **A**—Charge is conserved, meaning that while negative charge can neutralize positive charge, the net charge of a system cannot change. The net charge of state 1 is zero; thus, the net charge of state 2 must also be zero, regardless of whether the balls touched or not. State 2 has a net charge of −80 μC, not zero; thus, the claim is not reasonable due to this violation of charge conservation.

16. **C**—You want the wavelength of the sound wave, not the wave on the guitar. Choices A and B both correctly determine the wavelength of a wave on a guitar string, but these are not the waves you are looking for. Choice D does discuss the sound wave in air, but when is the last time you visually "saw" a sound wave passing you in the air, let alone saw any peaks that you could practically measure with a meterstick? So do the experiment in choice C. Going from one resonance to the next adds half a wavelength to the standing wave in the pipe, regardless of whether the pipe is open or closed.

17. **B**—You can look at this two ways. The hard way is to consider the spring energy gained and the gravitational energy lost in stretching the spring the additional 0.10 m separately. The block-earth system loses mgh = (0.20 kg)(10 N/kg)(0.10 m) = 0.20 J of gravitational energy; but the block-spring system gains $\frac{1}{2}kx_2^2 - \frac{1}{2}kx_1^2$ = {½(4 N/m)(0.60 m)^2 − ½(4 N/m)(0.50 m)^2} = 0.22 J of spring energy. Thus, the net work done on the block in pulling it the additional 0.10 m is 0.02 J. That's what is converted into the block's maximum kinetic energy.

You can also look at it the easy way. With a vertical spring, consider the block-earth-spring system as a whole. Define the hanging equilibrium as the zero of the whole system's potential energy; then the potential energy of the whole system can be written as $\frac{1}{2}kx^2$, where x is the distance from this hanging equilibrium position. That's ½(4 N/m)(0.10 m)^2 = 0.02 J.

18. **C**—The graph represents the speed of cart A, the one that's initially moving. So right before the collision, the vertical axis of the graph must be nonzero. Right after the collision, the vertical axis must quickly either decrease or perhaps become negative if the cart changed directions. That's what happens in the tenth of a second or so after the time labeled C.

19. **A**—Conservation of momentum requires that the total momentum of the two-cart system be the same before and after the collision. You already know cart A's speed and direction of motion before and after the collision by looking at the vertical axis of the graph; so the mass of cart A will give us cart A's momentum before

and after the collision. You know cart B has no momentum before the collision. But you need both cart B's mass AND its velocity after the collision to finish the momentum conservation calculation.

20. **D**—"Increase in the internal energy of the road-car system" is a fancy way of saying "work done by the car's brakes to stop the car." Without the brakes, the car would have gained mgh of mechanical energy in dropping the height h, giving it a total mechanical energy of $\frac{1}{2}mv^2 + mgh$. The brakes convert all of that mechanical energy to internal energy.

21. **B**—The equivalent resistance of the parallel resistors in circuit 1 is 12 kΩ; adding that to the 100 kΩ resistor gives a total resistance in circuit 1 of 112 kΩ. Circuit 2's parallel combination has an equivalent resistance of 23 kΩ, giving a total resistance in that circuit of 43 kΩ. By $V = IR$ applied to both circuits in their entirety with the same total voltage, the circuit with smaller total resistance will produce the larger current. That's circuit 2.

22. **A**—The motion after the push is finished is toward more positive x, so it is in the positive direction. The puck is slowing down. When an object slows down, its acceleration and its net force are in the direction opposite the motion. (Choice C would be correct at a position at which the student were still pushing the puck, but at $x = 1.0$ m, the student had already let go, so the puck was slowing down.)

23. **A**—The net force on the cart-object system is 2.0 N. Dividing 2.0 N by the acceleration gives the mass of the entire cart-object system, not just the mass of the cart; so subtract the 0.2 kg mass of the object from the whole system mass to get the cart mass. This is actually inertial mass, because it uses the equation $F_{net} = ma$; Newton's second law defines inertial mass as resistance to acceleration.

24. **D**—All of the diagrams have the object's weight correct. Any other force acting on the object must be provided by something in contact with that object. The only thing making contact with the object is the string, so add the force of the string on the object. That's it. The force of the

object on the string doesn't go on this diagram, because this diagram only includes forces acting ON the object, not exerted by the object.

25. **D**—Positions 1 and 4 show minimum particle displacement, so these are the nodes. The maximum amplitude occurs at the antinode, which is somewhere near the positions labeled 2 and 3. The particles are moving left and right in this longitudinal wave. The amplitude is twice the peak-to-peak displacement; since the particle displacement near the antinode is about the distance from 2 to 3, the amplitude is half that.

26. **A**—The rod starts from rest, so its final angular momentum is the same as its change in angular momentum. Although the equation $\omega = v/r$ is valid for a point object moving in a circle, it does not apply to a rotating rod; thus, choice B is wrong. Choice C gives the final angular momentum of the ball, which is not the same as the angular momentum change of the rod because the ball does NOT start from rest.

27. **C**—Choice C states the fundamental condition for angular momentum conservation, which is correct here. The ball does have angular momentum about the rod's center of mass before and after the collision, because its line of motion does not go through the rod's center of mass. Whether or not the collision is elastic has to do with conservation of mechanical energy, not angular or linear momentum.

28. **C**—Impulse is the area under a force-time graph. From $t = 8$ s to $t = 10$ s, the area is an approximate rectangle of 2 N times 2 s, giving 4 N·s. Add an approximate triangle from $t = 10$ s to $t = 11$ s, which has the area $\frac{1}{2}(2$ N$)(1$ s$) = 1$ N·s. That gives a total of 5 N·s.

29. **D**—The force of the engine on the rocket during this time is 2 N upward. The weight of the rocket is 1 N (that is, 0.1 kg times the gravitational field of 10 N/kg). So the net force is still upward during this time. Since the rocket was already moving upward, it will continue to move upward and speed up.

30. **D**—Impulse is change in momentum. The initial momentum was something like (0.5 kg)(1.6 m/s) = 0.8 N·s to the right. The cart came

to rest, changing its momentum by 0.8 N·s, then sped back up, again changing momentum by (0.5 kg)(2 m/s) = 1.0 N·s. Thus, the total momentum change is about 1.8 N·s.

31. **D**—The gravitational acceleration is given by GM/d^2, where d is the Space Shuttle's distance to Earth's center. You don't know values for G and M, nor do you need to know them. You do know that at Earth's surface, 6,400 km from the center, the gravitational acceleration is 9.8 m/s². To calculate using this equation at the height of the Space Shuttle, the numerator remains the same; however, the denominator increases from (6,400 km)² to (6,700 km)², a difference of about 8%. (Try it in your calculator if you don't believe me.) Thus, the gravitational acceleration will decrease by about 8%, giving choice D. (Yes, things seem "weightless" in the Space Shuttle. That's not because $g = 0$ there, but because everything inside the shuttle, including the shuttle itself, is in free fall, accelerating at 8.9 m/s/s toward the center of Earth.)

32. **B**—The scale reading is less than the man's weight; that means that the net force is downward. By Newton's second law, the person's acceleration is downward too. Downward acceleration means either moving down and speeding up, or moving up and slowing down. Only choice B works.

33. **C**—The weight of the textbook is GMm/d^2, where M and m are the masses of Earth and the textbook. These don't change on a mountain. The d term will change, because d represents the distance between the textbook and Earth's center. But that will change the denominator of the weight equation from (6,400 km)² to (6,406 km)²; in other words, not to the two digits expressed in the answers. No need to use the calculator. You can see that the choices require the weight to either stay the same, double, or be cut in half. The weight of the textbook remains 30 N.

34. **C**—In circular motion around a planet, the centripetal force is provided by gravity: $\dfrac{mv^2}{d} = G\dfrac{Mm}{d^2}$. Solving for the radius of the satellite's circular orbit, we get $d = \dfrac{GM}{v^2}$. The numerator of this

expression doesn't change, because the mass of the planet M and Newton's gravitation constant G don't change. The satellite's speed v is doubled. Since the v term is squared, that increases the denominator by a factor of 4. So the radius of orbit d is now ¼ as much as before.

35. **B**—Consider the cart alone, which as a single object has no internal energy or potential energy. Thus, any work done on the cart will change the cart's kinetic energy. The cart began with no kinetic energy at all. Earth increased the cart's kinetic energy by 0.30 J, as stated in the problem; the spring decreased the cart's kinetic energy by 0.20 J. A gain of 0.30 J and a loss of 0.20 J leaves 0.10 J of kinetic energy.

36. **B**—Conservation of linear momentum requires that the center of mass of the system continue to move to the right after the collision. The rotation will be about the combined rod-putty center of mass. To understand that, imagine if the putty were really heavy. Then after the collision, the rod would seem to rotate about the putty, because the center of mass of the rod-putty system would be essentially at the putty's location. In this case, you don't know whether the putty or the rod is more massive, but you do know that when the two objects stick together, they will rotate about wherever their combined center of mass is located.

37. **D**—No unbalanced forces act here other than the putty on the rod and the rod on the putty. (The weight of these objects is canceled by the normal force.) Thus, linear momentum is conserved. No torques act on the rod-putty system except those due to each other; thus, angular momentum is conserved. It's essentially a fact of physics that in a collision between two objects, both linear and angular momentum must be conserved.

38. **C**—The net torque on the rod must be zero, because the rod doesn't rotate. Therefore, whatever happens to Bob's torque must also happen to Tom's torque—these torques must cancel each other out. That eliminates choices A and B. The forces provided by Bob and Tom must add up to 500 N, the weight of the rod. Try doing two quick calculations: In the original case, Bob and Tom must each bear 250 N of weight and

are each 2 m from the midpoint, for 500 N·m of torque each. Now put Tom farther from the midpoint—say, 3 m away. For the torques to balance, $F_{bob}(2 \text{ m}) = F_{tom}(3 \text{ m})$. The only way to satisfy this equation and get both forces to add up to 500 N is to use 300 N for F_{bob} and 200 N for F_{tom}. Now, the torque provided by each is $(300 \text{ N})(2 \text{ m}) = 600$ N·m. The torque increased for both people.

39. **A**—The object is in equilibrium, so left forces equal right forces. Thus, the horizontal tensions must be the same in each rope. Rope A pulls at a steeper angle than rope B, but with the same amount of horizontal force as rope B. To get to that steeper angle, the vertical component of the tension in rope A must be larger than in rope B.

40. **A**—There's no indication that energy must be conserved in collision 1. However, momentum is always conserved in a collision. When block A bounces, its momentum has to change to zero and then change even more to go back the other way. Since block A changes momentum by more in collision I, block B must as well because conservation means that any momentum change by block A must be picked up by block B. Choices C and D are wrong because, among other things, they use conservation of momentum to draw conclusions about two separate collisions; momentum conservation means that total momentum remains the same before and after a single collision, not in all possible collisions.

41. **B**—The period of a mass-on-a-spring oscillator is $T = 2\pi\sqrt{\dfrac{m}{k}}$. The important part here is that the mass term is in the numerator—a larger mass means a larger period. More mass oscillates on the spring in collision II, so collision II has a greater period. Frequency is the inverse of the period, so the period is smaller in collision II.

42. **D**—The fundamental will have a node at one edge, an antinode at the other, and no nodes in between. The next allowable harmonic will have the same end conditions as the fundamental plus one additional node. A string fixed at one end and free to move at the other can only produce standing waves whose frequency is an odd multiple of the fundamental. So 30 Hz is the next

available frequency after the fundamental with one additional node.

43. **D**—Angular acceleration is net torque divided by rotational inertia, $\alpha = \dfrac{\tau_{net}}{I}$. Imagine each object has total mass of 2 kg. Begin by comparing objects A and B: To find the net torque on object A, assume the entire 20 N weight is concentrated at the dot representing the rod's center of mass. That's located 10 cm from the pivot, giving a net torque of 200 N·cm. For object B, consider the torques provided by each block separately. The right block provides a torque of $(10 \text{ N})(30 \text{ cm}) = 300$ N·cm clockwise; the left block provides a torque of $(10 \text{ N})(10 \text{ cm}) = 100$ N·cm counterclockwise. That makes the net torque 200 N·cm, the same as for object A. But object B has more rotational inertia, since its 2 kg of mass are concentrated farther away from the pivot than object A's mass. So the denominator of the angular acceleration equation is bigger for object B with the same numerator, which means A > B.

Now consider object C. It experiences less net torque than A and B. Calculate $(10 \text{ N})(30 \text{ cm}) = 300$ N·cm clockwise, and $(10 \text{ N})(20 \text{ cm}) = 200$ N·cm counterclockwise for a net torque of 100 N·cm. And object C has the same mass distributed even farther from the pivot point than either of the other two objects, giving C an even bigger rotational inertia. In the angular acceleration equation, object C gives a smaller numerator and a bigger denominator than object B, meaning C < B. Put it all together to get A > B > C.

44. **B**—Since the source is still moving at constant speed, the wave fronts will be equally spaced. However, since the source is moving, it will have traveled a bit with the wave before emitting the next wave. Thus, the wave fronts will be closer together. As an alternate explanation, a stationary observer at the right of the page would hear a higher frequency by the Doppler effect, meaning that more waves should pass the observer in one second; that requires wave fronts to be closer together.

45. **D**—Consider the system consisting of the objects and Earth, with the location of the $3m$ mass being the zero of gravitational energy. The

initial gravitational energy of the system is mgL. After the rotation, the final gravitational energy of the system is $3mgL$. That extra gravitational energy of $2mgL$ came from the work done on the system, meaning choice D. If you want instead to think of work on the objects as force times distance, remember that the force of Earth on the objects acts straight down, not along a circle. So the distance term to use here is just L, not πL.

Questions 46–50: Multiple-Correct Items (You must indicate both correct answers; no partial credit is awarded.)

46. **A and B**—Choice A uses the equation $F_{net} = ma$, which essentially defines inertial (as opposed to gravitational) mass. Choice B measures inertial mass, because it measures mass as resistance to the acceleration caused by the spring force. Choices C and D measure an object's behavior in a gravitational field; by definition, that's *gravitational*, not inertial, mass.

47. **C and D**—Though both speed divided by acceleration (as in choice A) and position divided by speed (as in choice B) give units of time, the time indicated has no relation to the period of the motion. The period is defined as the time for one complete cycle to happen. One complete cycle can be defined as the time for the object

to get back to its extreme position, as in choice D, or it can be defined as the time between the occurrences of maximum velocity in the same direction, as in choice C.

48. **B and C**—Elastic means that mechanical energy is conserved in the collision, so the answer is not choice A, which describes a loss of mechanical energy. Mechanical energy conservation requires that carts bounce, so choice B is correct, but choice D is not. Linear momentum is conserved in all collisions, elastic or not elastic.

49. **A and C**—The relevant equation here is $R = \rho \dfrac{L}{A}$. The length of the wire is the variable L, so choice C is correct. The diameter measurement (choice A) can be used with the formula for the area of a circle to get A, the cross-sectional area of the wire.

50. **B and C**—Change in angular momentum is $\tau \Delta t$, which means the area under a torque-time graph; thus, choice C is correct. Since this graph is linear, the average torque multiplied by the maximum time is the same thing as the area under the graph, so choice B is also correct. While choice D describes a calculation of angular momentum, it does not correctly give the *change* in angular momentum, so you don't know whether the object had an initial angular momentum or not.

Solutions: AP Physics 1 Practice Exam 2, Section I (Free-Response)

Obviously your solutions will not be word-for-word identical to what is written below. Award points for your answer as long as it contains the correct physics and as long as it does *not* contain incorrect physics.

Question 1

Part (a)
The resistor in series with the battery takes the largest current of any of the three resistors by Kirchoff's junction rule. By Ohm's law, $V = IR$, a resistor with both the largest current and the largest resistance will similarly take the largest voltage. So put the largest resistor R_3 in series with the battery, as shown.

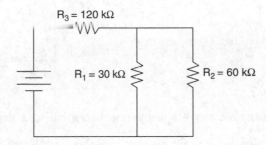

1 point for a correct diagram

1 point for a correct justification

Part (b)
Yes. Consider the equivalent circuit drawn, where the second resistor is the equivalent resistor of the parallel combination. These two resistors each take the same current by Kirchoff's junction rule; by Ohm's law, with the same current, the larger of the two takes the larger voltage. If the equivalent resistance of the parallel combination is greater than the resistance of the resistor in series with the battery, then the parallel combination will take more voltage than the first resistor. In this case, putting R_1 in series with the battery and the other two resistors in the parallel combination satisfies this condition; the equivalent resistance of R_2 and R_3 in parallel is greater than 30 kW.

1 point for comparing the resistance of the parallel combination to that of the series resistor

1 point for drawing the correct conclusion that the situation described is possible.

Part (c)

The total resistance of this circuit is the sum of the three resistors, 210 kΩ. The total current in the circuit is 12 V/210 kΩ = 57 µA (that is, 0.000057 A). Using that current in $V = IR$ for each resistor gives a voltage drop of 3.4 V, 6.9 V, and 1.7 V across each resistor in turn. So start the graph at 12 V. The voltage drops by 3.4 V after the 60 kΩ resistor, down to 8.6 V, as shown in the diagram. (The voltage does not change along the wire itself, which has essentially zero resistance compared to the resistors.) When we get to the 120 kΩ resistor, drop the voltage another 6.9 V, down to 1.7 V left. The voltage drops the rest of the way to zero after the 30 kΩ resistor, as required by Kirchoff's loop rule.

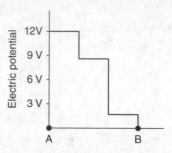

1 point for any graph having three distinct voltage drops ending at 0 V.

1 point for any graph that somehow indicates the largest voltage drop across the middle resistor, and the smallest voltage drop across the last resistor

1 point for any graph with three distinct horizontal segments representing the voltage in the wires not changing.

Question 2

Part (a)

The maximum coefficient of static friction is greater than the coefficient of kinetic friction. The graph shows the maximum force of static friction, when the block starts moving, to be 1.7 N; when the block is moving, the friction force drops to 1.0 N, which is appropriate for the lower kinetic friction force.

1 point for correctly explaining the difference between static and kinetic friction in this case.

Part (b)

(i) Impulse is the area under a force-time graph. The graph has an average force of about 1.0 N, and a time interval of about 2 s (from $t = 2$ s to $t = 4$ s). That's an impulse of 2 N·s.

1 point for using the area under the F vs. t graph

1 point for making reasonable estimates of time and force from the graph

(ii) The student is incorrect. While it is true that the net impulse on an object is equal to the object's change in momentum, the graph shown does not include the net force. The graph shows the force of the string only. If you were to subtract the impulse provided by the friction force, you would get zero net impulse and thus zero change in momentum.

1 point for discussing the difference between the force of the string and the net force on the cart

1 point for including no incorrect statements

Part (c)

The student could use a set of toy carts, each of which moves at a different constant speed. The student should connect the carts to the block via the force probe and produce force-time graphs for each cart as the cart moves across the table. The constant speed traveled by each cart can be measured by placing a sonic motion detector in front of the cart.

3 points for a complete and correct description. Of these points, 1 or 2 could be holistically awarded for a partially complete and/or partially correct description.

Part (d)

The student should make a plot of speed on the vertical axis versus the force of kinetic friction on the horizontal axis. The speed of each cart would be determined by the slope of the position-time graph produced by the sonic motion detector. The force of kinetic friction would be measured by the force probe.

The block is the same mass each time, which means it always experiences the same normal force; the coefficient of friction is the friction force divided by the normal force. So if the coefficient of friction changes, the force probe reading would change as well.

If this graph is horizontal, then the force of friction and the coefficient of friction do not change with speed. If the graph is sloped, then we can conclude that the coefficient of friction *does* change with speed.

1 point for using a graph with a significant number of data points, or perhaps statistical analysis of multiple trials.

1 point for correctly relating the friction coefficient to the force probe reading with constant F_n.

2 points for a complete and correct analysis. One of these points can be holistically awarded for a partially complete and/or partially correct analysis.

Question 3

Part (a)

1 point: As the hanging object m falls, it speeds up. Its acceleration is therefore downward, and so is the net force acting on it. To get a downward net force, the downward gravitational force mg must be greater than the rope's tension.

Part (b)

4 points:

- **1 point** for using an applicable fundamental relationship such as $\alpha = \frac{\tau_{net}}{I}$ (or $a = r\alpha$.)
- **1 point** for a correct discussion of how to get τ_{net} (or α)
- **1 point** for a correct discussion of how to estimate I (or a)
- **1 point** for a correct conclusion to answer the question

The angular acceleration can be found by dividing the net torque on the device by its rotational inertia. The rotational inertia I of the device can be calculated based on the assumption that the pipe and the stem do not contribute; the device consists of two pointed masses, each a distance of $L/2$ from the center of rotation. The net torque on the device is Tr, where r is the radius of the support.

The only unknown information in all of this is the radius of the support r. So **no, the angular acceleration cannot be calculated** with the information provided unless r is measured with calipers.

(*Alternate solution:* The angular acceleration of the device α is related to the linear acceleration a of the falling mass by $a = r\alpha$. The radius of the support r must be measured with calipers. To get a, the distance the mass falls from rest d can be measured with a meterstick, and the time t to fall that distance can be measured with a stopwatch. Then kinematics can be used to calculate a. So **no, the angular acceleration cannot be calculated** with the information given here.)

Part (c)

1 point for clearly defining and correctly using any new variables

1 point for a correct expression for I (or a)

1 point for a correct final answer

As described in part (b), let r be the radius of the support.

By Newton's second law for rotation, $\alpha = \frac{\tau_{net}}{I}$.

Using the net torque and rotational inertia explained in part (b), $\alpha = \dfrac{Tr}{2M\left(\dfrac{L}{2}\right)^2}$.

(*Alternate solution, as described in part (b):* $\alpha = \dfrac{a}{r}$.)

Falling from rest means zero initial velocity, so $d = \frac{1}{2}at^2$, with d and t defined as shown. Algebraically, this makes $a = \dfrac{2d}{t^2}$.

Combining these two statements gives $\alpha = \dfrac{2d}{rt^2}$.)

Part (d)

1 point for describing the quantity in the equation in (c) that would be affected by the new mass

1 point for describing with reference to the equation how that change would affect the acceleration

While the tension in the hanging rope is not equal to the weight of the hanging object, increasing the hanging object's mass would increase the tension in the rope T. Since T is in the numerator of the equation for angular acceleration, and since the parameters in the denominator are unchanged, the angular acceleration would increase.

(*Alternate solution:* We would be able to measure that the hanging object falls the same distance in less time. Note that this is NOT because "heavier objects fall faster"; this is really a consequence of the reasoning already given, that the tension in the rope increases without changing the properties of the device. With a smaller t in the denominator and other variables unchanged, the angular acceleration would increase.)

Part (e)

2 points for a complete and correct solution. One of these points can be awarded for a partially complete or partially correct solution.

Using a calculational approach: You assumed that the rotational inertia of the device was wholly due to the two rocks as point objects, $2M(L/2)^2$. The rotational inertia of a cylinder rotating horizontally and vertically can be looked up, and the mass of the pipe and support can be measured. You could calculate the additional rotational inertia provided by the cylinder and support. If this additional rotational inertia is substantially less than $2M(L/2)^2$ such that the calculation of α would still come out approximately the same, then this additional inertia is negligible.

 Using an experimental approach: Measure the angular acceleration of the device directly. This can be done with frame-by-frame video analysis or with a photogate set to measure the increase in angular velocity. If the angular acceleration measured matches that predicted by the equation in (c), then the assumption that the rotational inertia is due wholly to the point masses is reasonable. However, if the angular acceleration is measured to be noticeably smaller than that predicted in (c), then the rotational inertia of the pipe and support do contribute meaningfully to the calculation and are not negligible.

Question 4

Part (a)

(i) **1 point** for calculating the spring constant

 1 point for using the calculated spring constant in a correct equation to determine the work done

The spring constant of the spring can be determined by procedure A. The rock applies a 4 N force on the spring, compressing it 0.05 m. By $F = kx$, that gives a spring constant k of $(4 \text{ N})/(0.05 \text{ m}) = 80 \text{ N/m}$.

 The potential energy of the spring-block-Earth system is just $\frac{1}{2}kx^2$, where x is the distance from the position of the block after procedure A. (If you were talking about just the spring-block system, you would use the distance from the undisturbed position, but then you would have to consider the work done on the block-Earth system separately.) So in procedure B, the block-Earth system gains $\frac{1}{2}(80 \text{ N/m})(0.05 \text{ m})^2$ of potential energy, which is 0.10 J. That's how much work was done by the student.

(ii) **1 point** for correctly justifying the use of data from procedure A to get the spring constant

 1 point for justifying use of $\frac{1}{2}kx^2$ as a change in potential energy of the correct system

See above.

Part (b)

1 point for recognizing that the same potential energy is available to be converted to KE

1 point for words or equations showing that the mass only shows up in the denominator of an expression for the height

1 point for using an equation or conservation of energy reasoning to get ½h.

The new spring-block-Earth system stores the same 0.10 J of potential energy: $\frac{1}{2}kx^2$, where x is the 0.05 m distance from the position of the block after procedure C. That 0.10 J is converted into kinetic energy, then to purely gravitational energy. Gravitational energy is mgh; the new height is $h_{new} = 0.10 \text{ J}/mg$. Since this new mass is twice as much as before, and since height is in the denominator, the new height is half as much as h.

Question 5

- **1 point** for a clear, coherent paragraph-long explanation, one in which a reader can immediately and easily understand the proposed procedure
- **1 point** for using a practical procedure, i.e., one that *could* easily be performed in a lab, even if the spirit of "no equipment whatsoever" was violated, or even if it wouldn't allow determination of the required frequency
- **1 point** for using a procedure that would work in principle to find the frequency, whether practical or not.
- **1 point** for an experiment that would work and meets the spirit of "no equipment whatsoever"
- **2 points** for clearly and correctly explaining the fundamental physics principles underlying the method
- **1 point** for connecting those physics principles to the specific method used.

Since the speakers are of slightly different frequencies, you should be able to hear beats. The beats occur because the speakers' notes alternately interfere constructively and then destructively. The frequency of the beats is equal to the difference in frequencies.

Start by standing between the speakers and hearing the beats. Estimate the number of beats that happen in one second, perhaps by having another student count out 10 s while you count out the total beats in those 10 s. Call the number of beats in each second f. Then f is the difference between the frequencies; the right-hand speaker could be either (350 Hz $+ f$) or (350 Hz $- f$).

Finally, listen to each note separately to determine which has the higher pitch. Pitch of a sound is related to its frequency. That will allow you to determine which equation applies.

Scoring the Practice Exams

First, please understand that no one outside the offices of the Educational Testing Service has any clue whatsoever exactly how much credit will be necessary to earn a 5, 4, 3, etc. Historically on the AP Physics B exam, it took about 65 percent of the points to earn a 5; 50 percent to earn a 4; and 35 percent to earn a 3. These cutoff scores change slightly every year—the goal is to maintain consistency across the years as to the standard of performance represented by each score.

This new exam will likely have totally different score cutoffs. The problem is, anyone who says he or she can tell you exactly *how* those cutoff scores will change is either breaking confidentiality by giving you proprietary ETS information; or, more likely, selling you a pile of steaming horse doings.

I'm going to take the middle ground—I'm going to give a scoring chart with the same approximate cutoffs as in previous years. This is the chart I'll use in my classes for now, until I see an authentic chart eventually. Use it at your own peril.

Multiple-Choice Raw Score: Number Correct_____(50 points maximum)

Free Response: Problem 1_____(12 points maximum)
 Problem 2_____(12 points maximum)

 Problem 3_____(7 points maximum)

 Problem 4_____(7 points maximum)

 Problem 5_____(7 points maximum)

Free response total:_____(45 points maximum)

The final score is equal to $(1.11 \times$ the free response score$) + ($the multiple choice score$)$

Total score:_____(100 points maximum)

Approximate Score Conversion Chart (Only a Guesstimate, See Above)

Raw Score	AP Grade
65–100	5
50–64	4
35–49	3
21–34	2
0–20	1

Appendixes

Table of Information
The Pantheon of Pizza

TABLE OF INFORMATION

You will be given this information as part of the AP Physics 1 Exam. It's worth checking out the official version of the table at Collegeboard.org—they may use slightly different symbols and layout than you see here.

Constants

Gravitational field at Earth's surface	g	10 N/kg
Universal Gravitation constant	G	6.7×10^{-11} N·m²/kg²
Coulomb's law constant	k	9.0×10^{9} N·m²/C²
Elementary charge	e	1.6×10^{-19} C
Speed of light in a vacuum	c	3.0×10^{8} m/s
Mass of an electron	m_e	9.1×10^{-31} kg
Mass of a proton	m_p	1.7×10^{-27} kg

Trigonometry

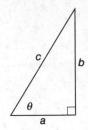

$\sin \theta = b/c$

$\cos \theta = a/c$

$\tan \theta = b/a$

Mechanics Equations

$v_f = v_0 + at$

$\Delta x = v_0 t + \dfrac{1}{2} at^2$

$v_f^2 = v_0^2 + 2a\Delta x$

$a = \dfrac{F_{net}}{m}$

$F_f = \mu F_n$

$p = mv$

$\Delta p = F \cdot \Delta t$

$KE = \dfrac{1}{2} mv^2$

$$KE_{\text{rotational}} = \frac{1}{2} I\omega^2$$

$$PE = mgh$$

$$PE = -G\frac{M_1 M_2}{d}$$

$$PE = \frac{1}{2}kx^2$$

$$W = F\Delta x_{\parallel}$$

$$W_{\text{NC}} = (\Delta KE) + (\Delta PE)$$

$$a_c = \frac{v^2}{r}$$

$$I = \sum mr^2$$

$$\tau = Fd_{\perp}$$

$$\tau_{net} = I\alpha$$

$$\Delta L = \tau \cdot \Delta t$$

$$w = mg$$

$$g = G\frac{M}{d^2}$$

$$F = \frac{Gm_1 m_2}{d^2}.$$

Electricity Equations

$$F = k\frac{Q_1 Q_2}{d^2}$$

$$R = \frac{\rho L}{A}$$

$$V = IR$$

$$P = IV$$

Wave and Simple Harmonic Motion Equations

$$T = 2\pi \frac{\sqrt{m}}{\sqrt{k}}$$

$$F_s = kx$$

$$T = 2\pi \frac{\sqrt{L}}{\sqrt{g}}$$

$$f_1 = \frac{v}{2L}$$

$$f_1 = \frac{v}{4L}$$

$$v = \lambda f$$

THE PANTHEON OF PIZZA

Pizza is the traditional food of the physics study group. Why? Probably because it's widely available, relatively inexpensive, easily shareable, and doesn't cause arguments the way "let's order bean curd" might.

If you have not yet experienced the late-night physics group study session, you should. Physics is more fun with friends than alone, and you learn more productively with other people around. Ideally, you'll find a mix of people in which sometimes they are explaining things to you, but sometimes you are explaining things to them. Explaining physics to friends is the absolute best way to cement your own knowledge.

But if you don't already have a regular study group, how do you go about creating one? Use pizza as bait. "Hey, let's get together in my mom's basement to do the problem set" is like a party invitation from Bill Nye the science guy. But, "Hey, we're ordering seven large pizzas with extra cheese and a variety of toppings, why don't you come by and do your problem set with us?" sounds more like you're headed to *Encore* on the Vegas Strip.[1]

Over the years I've eaten enough pizza to fill several dozen dumpsters—and dumpsters have been an appropriate receptacle for much of that pizza. Given the choice between a five-star restaurant and a pizza place, I'd usually choose the five stars. Usually. I know of four—just four—pizza places I would prefer to anything recommended by Squilliam Fancyson.

These four make up the Pantheon of Pizza.

Please understand the rules of access to the Pantheon:

1. I must have eaten at a member restaurant at least twice. This unfortunately rules out the heavenly *Pepper's Pizza* in Chapel Hill, North Carolina.[2]
2. I must have such an affinity for their pepperoni-and-extra-cheese pizza that my mouth waters upon the mere mention of a potential visit to the restaurant.

That's it. It's my pantheon, so it's my choice who gets in.

That said, please do send your own corrections, additions, oversights, etc. You can contact me via Woodberry Forest School. If you make a good enough case for a particular pizza place possibly joining the Pantheon, I may attempt to make a pilgrimage.

The Pantheon

4. Broadway Joe's Pizza, Riverdale, New York. This tiny shop below the #1 line train station in the North Bronx has everything you could ask for in a New York pizzeria— street noise, no air conditioning but instead a fan running all summer, the Yankees game on the television, and Broadway Joe himself behind the counter. Okay, I'm sure that there are hundreds of such places throughout New York City, all of which probably have

---

[1]Sorry. I certainly do *not* intend to dis Bill Nye the Science Guy. He is demonstrably cool.
[2]Now, alas, I hear they've closed down for good. Sigh.

tremendous pizza. But Broadway Joe's is the one I walk to every year during the AP Physics Teachers' Summer Institute that I run at Manhattan College. Bonus points to Mr. Joe for recognizing me each year: "Hey, you're the teacher who wants a small[3] pepperoni and extra cheese." Ten minutes later, out pops the classic New York–style pie with deciliters of cheese piled on top of a foldable crust. I can never finish the small by myself, but I so, so want to.

3. Thyme Market, Culpeper, Virginia. When I moved to central Virginia, I initially despaired at the food choices. But then we discovered Pizza Monday at Thyme Market. Even an unjuiced Alex Rodriguez[4] could knock a baseball across the length of Culpeper's Main Street, but it contains the heavenly brick oven from which its $5 pies spring forth each Monday, plus a lot of money for toppings, plus another $5 if you come on a day other than Monday. But it's well worth the cash and the trip. This is one of the few pizza places ever in which "extra cheese" provides a bit too much gooeyness. The pepperoni itself is the best of any in the Pantheon—just the right size, on top of the cheese, a bit of thickness to it, and baked until the edges begin to get crispy. While you're in the restaurant, try a bite of the "Culpeper Crack" branded cheese spread that's always available to sample. The pizza may be only $5 per pie, but you'll spend an order of magnitude more than that after you buy up multiple tubs of the Crack to take home.

2. Big Ed's Pizza, Oak Ridge, Tennessee. I encountered Big Ed's in conjunction with the United States Invitational Young Physicists Tournament, which was held in Oak Ridge for several years due to the presence of Oak Ridge National Laboratory. My friend and fellow physics teacher Peggy insisted that it was worth waiting in the crowd outside the door for a table, and she was right. The pizza was, of course, fabulous: New York–style foldable crust, with plenty of cheese and a multitude thereof of pepperonis. What sold me on Big Ed's, more so than even the T-shirts with a cartoony drawing of Big Ed himself, was the Kneeling Bench. The kitchen is separated from the dining area by a high wooden façade. But in the middle of the façade are two holes, with benches underneath. I was instructed to kneel on a bench, cup my hands communion-style, and put them through the hole. Lo, a generous portion of shredded mozzarella was placed in my hands by unknown beneficiaries. I had to go through this ritual a second time—I probably had as much cheese from my trips to the Kneeling Bench as from the pizza itself.

1. Langel's Pizza, Highland, Indiana. Most people who sample Chicago-style deep-dish pizza go to the big chains that have sprouted up across the Chicagoland area. Burrito Girl[5] grew up in a small suburb in northwest Indiana, and so she is well aware of the famous, fancy chains. Yet the first pizza place that Burrito Girl took me to consisted of about six booths sandwiched between an exotic reptiles store and a sports bar. I ordered, and I endured the requisite progression of helpfulness, skepticism, and then outright horror that waitresses in Chicago pizza places bestow upon me when I order extra cheese. Yes, I want extra cheese, even though the pizza is stuffed with seemingly an entire cow-day's worth of cheese already. Really. I've done this before, and lived to tell the tale. Please?

At Langel's, the extra cheese oozes and stretches beyond the mere constraints of slices. It takes a full 20 minutes before the cheese is congealed enough to hold the shape in which

---

[3]Don't be deceived. A "small" pie at Broadway Joe's could last for three straight late nights of Minecraft. Just one extra-large could sustain the entire rat population along the banks of the East River.

[4]… if such a thing exists

[5]My wife and sidekick, also known as the mild-mannered Shari.

you cut it. But it's the sauce that makes Langel's the best pizza in the known universe. This deep-dish pizza does not come in layers, but rather mixed all about, which means that the sauce can be appreciated throughout every bite. The pepperoni is fine, but I actually recommend just getting a pure extra-cheese pie. You'll have enough for lunch right now, dinner tonight, and probably breakfast tomorrow. Too bad they don't deliver within a 1,200 km radius.